# Biosystems & Biorobotics

Volume 26

**Series Editor**

Eugenio Guglielmelli, Laboratory of Biomedical Robotics, Campus Bio-Medico University of Rome, Rome, Romania

The BIOSYSTEMS & BIOROBOTICS (BioSysRob) series publishes the latest research developments in three main areas: 1) understanding biological systems from a bioengineering point of view, i.e. the study of biosystems by exploiting engineering methods and tools to unveil their functioning principles and unrivalled performance; 2) design and development of biologically inspired machines and systems to be used for different purposes and in a variety of application contexts. In particular, the series welcomes contributions on novel design approaches, methods and tools as well as case studies on specific bio-inspired systems; 3) design and developments of nano-, micro-, macro- devices and systems for biomedical applications, i.e. technologies that can improve modern healthcare and welfare by enabling novel solutions for prevention, diagnosis, surgery, prosthetics, rehabilitation and independent living. On one side, the series focuses on recent methods and technologies which allow multi-scale, multi-physics, high-resolution analysis and modeling of biological systems. A special emphasis on this side is given to the use of mechatronic and robotic systems as a tool for basic research in biology. On the other side, the series authoritatively reports on current theoretical and experimental challenges and developments related to the "biomechatronic" design of novel biorobotic machines. A special emphasis on this side is given to human-machine interaction and interfacing, and also to the ethical and social implications of this emerging research area, as key challenges for the acceptability and sustainability of biorobotics technology. The main target of the series are engineers interested in biology and medicine, and specifically bioengineers and bioroboticists. Volume published in the series comprise monographs, edited volumes, lecture notes, as well as selected conference proceedings and PhD theses. The series also publishes books purposely devoted to support education in bioengineering, biomedical engineering, biomechatronics and biorobotics at graduate and post-graduate levels.

Indexed by SCOPUS and Springerlink. The books of the series are submitted for indexing to Web of Science.

More information about this series at http://www.springer.com/series/10421

Irfan Hussain · Domenico Prattichizzo

# Augmenting Human Manipulation Abilities with Supernumerary Robotic Limbs

Irfan Hussain
Department of Mechanical Engineering
Khalifa University Center for Autonomous
Robotic Systems (KUCARS)
Khalifa University of Science
and Technology
Abu Dhabi, United Arab Emirates

Domenico Prattichizzo
Department of Information Engineering
Università degli Studi di Siena
Siena, Italy

ISSN 2195-3562 ISSN 2195-3570 (electronic)
Biosystems & Biorobotics
ISBN 978-3-030-52004-5 ISBN 978-3-030-52002-1 (eBook)
https://doi.org/10.1007/978-3-030-52002-1



This Springer imprint is published by the registered company Springer Nature Switzerland AG
The registered company address is: Gewerbestrasse 11, 6330 Cham, Switzerland

# Preface

Supernumerary robotic limbs are a recently introduced class of wearable robots that, differently from traditional prostheses and exoskeletons, aim at adding extra arms, legs, or fingers to the human user, rather than substituting or enhancing the natural ones. The additional robotic limbs could let the human to augment their abilities and could give support in everyday tasks. The advantage of using supernumerary robotic devices is twofold. From one side, this addition can enable humans to augment their capabilities, see Fig. 1a. In the other side, extra limbs can compensate the missing abilities of impaired limbs, e.g., in case of stroke patients as shown in Fig. 1b. These wearable robots are aimed to augment not only the strength and the precision of the human users, but also their range of skills and interactions with the environment. Since supernumerary robotic limbs are supposed to closely interact and perform actions in synergy with the human limbs, the control principles of robotic devices and their structure should have similar behavior as human's ones.

(a) Bonilla, B.L. *et al.* [1] (b) Hussain.I *et al.* [2]

**Fig. 1** Supernumerary robotic limbs: the extra robotic arms and fingers to augment and compensate the abilities of natural limbs

The objective of this book is to lay the foundations of a novel approach to human–robot interaction, exploiting the emerging field of supernumerary robotic limbs in enhancing and compensating the capabilities of natural limbs. The idea of supernumerary limbs is ground breaking because the concept is novel and cannot be framed in any of the current areas of robotics. For example, the extra finger(s) proposed in this book augments the manipulation capability of the paretic limb without relying on the subject's skeletal structure (in contrast to exoskeletons), making anatomical variation and motion restriction a minor issue. Moreover, the extra finger is highly wearable and can be even transformed into a bracelet when not in use. The challenge lies not only in the design and development of these devices but also their integration with human body through the sensorimotor interfaces to enable user feeling the robot effectively, as a sort of natural extension of their sensorimotor abilities.

In this direction, this book contributes in proposing new generation of supernumerary robotic fingers along with their customized control interfaces that can be used in augmenting and compensating the manipulation abilities of human. The guiding principles of the robotic devices design are safety, wearability, ergonomics, and user comfort. Throughout the book the other synonyms used for supernumerary robotic finger(s) are extra robotic finger(s) or sixth finger.

Firstly, we investigate how to enhance the capabilities of the human hand by means of wearable robotic fingers. Adding wearable robotic fingers could give humans the possibility to manipulate objects in a more efficient way, enhancing our hand grasping dexterity and ability. In this regard, we have designed and developed a family of extra fingers ranging from fully actuated to underactuated. Together with the design issues related to portability and wearability of the devices, another critical aspect is integrating the motion of the extra fingers with the human hand. We proposed three possible control strategies along with their suitable robotic devices in terms of actuation and sensing capabilities. The first control strategy is based on mapping algorithm able to transfer to the extra fingers a part or the whole motion of the human hand. The second one is based on wearable sensorimotor haptic interfaces which enable the user to control not only the motion of robotic fingers but also perceive the information related to robotic finger status in terms of contact/no contact with the grasped object and in terms of force exerted by the device. The third one is Electromyographic (EMG) control interfaces to control motion and joints compliance of a supernumerary robotic finger.

Secondly, we explore the real potential of supernumerary robotic fingers as new generation of wearable assistive technology for stroke patients. Stroke and amputation are the two main causes of disabilities of the upper limb. While the scientific community made remarkable advancements in robotic prostheses for upper limb amputees [3], only few results are available to compensate for missing manipulation abilities in subjects with paretic upper limbs, such as subjects suffering from chronic stroke or other pathologies. According to the World Health Organization, stroke is the disease which leads to most of deaths and high disability problems. Approximately 60% of stroke survivors suffer from some form of loss of sensory and/or motor function of the hand [4]. For people with paretic upper limb, most of the attention of the community has been focused on exoskeletons [5] which are very difficult to use because of their limitation in accommodating the subjects anatomical variations due to impairment. Moreover, the poor wearability, in terms of weight and size make them difficult to use in Activities of Daily Living (ADL). One of the biggest challenges of rehabilitation and assistive engineering is to develop technology to practice intense movement training at home [6]. The creation of a functional grasp by means of the extra fingers enables patients to execute task-oriented grasp and release exercises and practice intensively using repetitive movements. Supernumerary robotic fingers can increase patients' performances, with a focus on objects manipulation, thereby improving their independence in ADL, and simultaneously decreasing erroneous compensatory motor strategies for solving everyday tasks. The idea of wearable supernumerary limbs as assistive devices is different in nature than other approaches used in rehabilitation and assistive robotics. Extra limbs will provide novel opportunities to recover missing abilities, resulting in improvements of patients' quality of life.

We also present the combination between the supernumerary robotic finger and the arm support that can be used during the rehabilitation phase when the arm is potentially able to recover its functionality, but the hand is still not able to perform a grasp due to the lack of an efficient thumb opposition. The overall system also acts as a motivation tool for the patients to perform task-oriented rehabilitation activities, e.g., box and block, Peg and hole, and Frenchay arm test. With the aid of proposed system, the patient can closely simulate the desired motion with the non-functional arm for rehabilitation.

Although mainly the current hope is to use the supernumerary robotic fingers for compensation, we have good expectation in using the device to rehabilitate at least for the arm. Moreover, we also believe that extra fingers can play a role even in hand rehabilitation. Patients with hemiparesis often have limited functionality in the left or right hand. The standard therapeutic approach requires the patient to attempt to make use of the weak hand even though it is not functionally capable, which can result in feelings of frustration. The aim is to provide patients with a sense of purpose and accomplishment during ADL training, even during the early phase of treatment when the task can only be partially completed. We hope that proposed devices can facilitate therapist-guided ADL training and encourage patients to continue exercising the affected limb.

From the neuroscientific point of view, the development of this type of devices opens a series of interesting questions on how extra limbs are perceived by human cognitive system that needs to be investigated.

The next chapters will provide the details of relevant contributions I made in the field of supernumerary robotic limbs. In particular, Chap. 1 introduces the supernumerary robotic fingers and state of the art. Chapter 2 investigates how the extra robotic fingers can augment the abilities of human hand to increase its work space and manipulation abilities. We propose the different design guidelines to realize a robotic extra finger for human grasping enhancement. Such guidelines were followed for the realization of three prototypes obtained using rapid prototyping techniques, i.e., a 3D printer and an open hardware development platform. Both fully actuated and underactuated solutions have been explored. We presented mapping algorithm able to transfer to the extra fingers a part or the whole motion of the human hand. Chapter 3 presents how the supernumerary robotic finger can be used to compensate the missing abilities on impaired upper limb in chronic stroke patients. The robotic finger and paretic limb act as two parts of a gripper to grasp and hold the object. Chapter 4 presents the design, analysis, fabrication, experimental characterization, and evaluation of two prototypes of soft supernumerary robotic fingers that can be used as grasp compensatory devices for hemiparetic upper limb. In Chap. 5, the wearable sensory motor interfaces for supernumerary robotic fingers are presented. In particular, two kinds of interfaces, namely, "vibrotactile ring" and hRing are proposed. The human user is able to control the motion of the robotic finger through a switch placed on the ring, while being provided with cutaneous feedback about the forces exerted by the robotic finger on the environment. In Chap. 6, two kinds of Electromyographic (EMG) control interfaces for supernumerary robotic fingers are presented. In particular, frontalis muscle cap and arm EMG interface. The former is more suitable for underactuated soft fingers while the later is developed to control the motion and compliance of fully actuated robotic finger. In frontalis muscle cap, the electrodes and acquisition boards are embedded in cap which allows the user to control the device motion through wireless communication by contracting the frontalis muscle. Chapter 7 presents the combination of soft extra finger with the arm support to provide the needed assistance during paretic upper limb rehabilitation involving both grasping and arm mobility to solve task-oriented activities. The book concludes with Chap. 8 by addressing how the supernumerary robotic fingers are providing novel opportunities to recover missing abilities of stroke patients, resulting in improvements of patients' quality of life. Although until now the supernumerary robotic fingers are mainly being used for grasping compensation, there is a good expectation in using these devices to rehabilitation. From the neuroscientific point of view, the development of this type of devices opens a series of interesting questions on how the supernumerary limbs are perceived by human cognitive system that are still in process of investigation.

Abu Dhabi, United Arab Emirates — Irfan Hussain
Siena, Italy — Domenico Prattichizzo

# Contents

# About the Authors

**Dr. Irfan Hussain** is an Assistant Professor at the Mechanical Engineering Department at Khalifa University (KU), Abu Dhabi, UAE. He holds a Ph.D. degree in Robotics from the University of Siena, Italy. His research interests include supernumerary (extra) robotic limbs, medical robotics, soft robotic hands, complaint manipulators, exoskeletons, and prostheses.

**Prof. Domenico Prattichizzo** is a Professor of Robotics at the University of Siena, Siena, Italy, and since 2009 has been a Scientific Consultant at Istituto Italiano di Tecnologia, Genova, Italy. In 1994, he was a Visiting Scientist at the MIT AI Lab. He received his Ph.D. degree in Robotics and Automation from the University of Pisa, Italy, in 1995. His research interests include supernumerary (extra) robotic limbs, haptics, grasping, visual servoing, mobile robotics, and geometric control.

# About the Authors

# Chapter 1
# Introduction

*There is nothing comparable to the human hand outside nature. We can land men on the moon, but, for all our mechanical and electronic wizardry, we cannot reproduce an artificial fore-finger that can feel as well as beckon.*
*John Napier*

Over many centuries the human has been extending their physical, cognitive, and social abilities in all activities. Every major technological breakthrough in the history made the human more capable, which changed the life style and the society. These changes have been many orders of magnitude faster than evolutional changes. Robotic applications have rapidly grown from classical industrial applications with repetitive tasks to applications with close human-robot interaction. In particular, assistive robotics has gained an increasing attention in the last decades, see [7].

Among all the robotic solutions, wearable robotics is one of the potential enabling technology to improve the humans quality of life. Wearable robotics is increasingly attracting research activity. Wearable robots are expected to work very closely, to interact and collaborate with people in an intelligent environment [8]. Prosthetic devices, exoskeletons and powered suites, wearable sensors and implantable devices are just a few examples of the wearable robots that have been developed within the robotics community in recent years [9, 10]. Although, two types of wearable robots, prostheses and exoskeletons, have been studied extensively, yet numerous technical challenges still exist, particularly for the upper extremities. Moreover, these devices are primarily developed for compensating specific disabilities for a certain patient population.

These are mainly used in substitution of lost limbs or for human limb rehabilitation or to amplify strength [5]. In particular, a prosthesis is an artificial device that replaces a missing body part, which may be lost through trauma, disease, or congenital conditions as shown in Fig. 1.1a. Prostheses are beneficial in improving



I. Hussain and D. Prattichizzo, *Augmenting Human Manipulation Abilities with Supernumerary Robotic Limbs*, Biosystems & Biorobotics 26,
https://doi.org/10.1007/978-3-030-52002-1_1

(a) Hand Prosthesis [11]

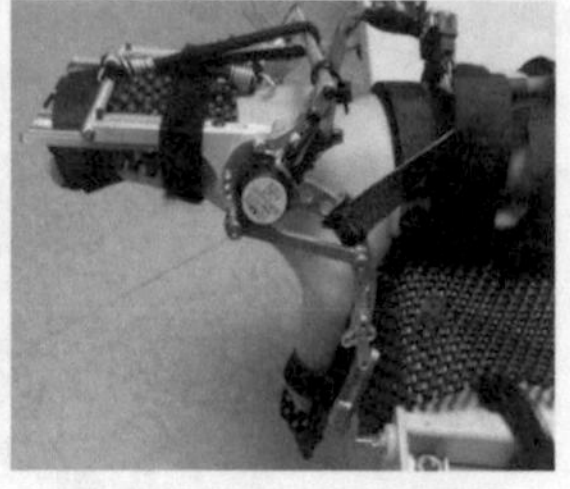

(b) Hand Exoskeleton [12]

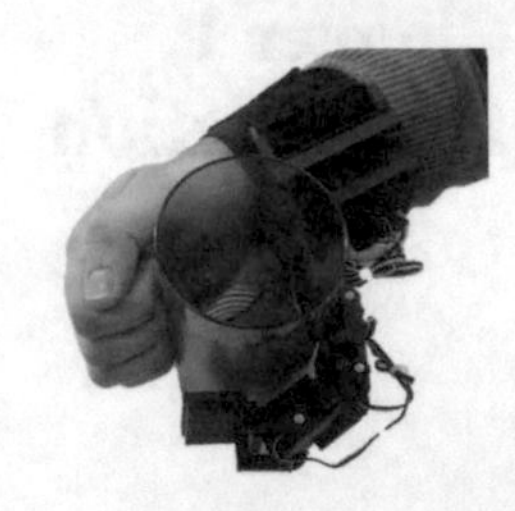

(c) Supernumerary robotic finger [2]

**Fig. 1.1** Examples of hand prosthesis, exoskeleton and supernumerary robotic finger are shown: prostheses are used in substitution of lost limbs while exoskeleton are suitable for human limb rehabilitation or to amplify strength. The supernumerary robotic fingers are addition to human limbs to supplement and augment their abilities without substituting them

the quality of life of amputees, but they are not suitable for hemiplegic and hemiparetic patients since the impaired arms and hands are still physically attached to their body. An exoskeleton is a wearable robot with joints and links corresponding to those of the human body [9] (see, Fig. 1.1b). Exoskeletons are robotic external structures, used to amplify strength and endurance, giving the ability to perform else impossible tasks (e.g. carrying heavy loads while working on a hard terrain), or to help rehabilitation of people who lost some motion functionalities. However, the technological development is reaching a turning point where the wearers of the devices can exceed the performance of average people. There are already cases where a handicapped athlete can run faster and a longer distance with specific augmentation devices. The technology is ready to be extended from compensation to augmentation and enhancement.

Apart from traditional prosthesis and exoskeletons, a very challenging research direction is to add robotic limbs to human, rather than substituting or enhancing them (see, Fig. 1.1c). This addition could let the human augment their abilities and could give support in everyday tasks. In [13–18], the authors presented the Supernumerary Robotic Limbs (SRL), a wearable robot designed to assist human with additional arms attached to the wearer's body. Such devices, further investigated in [19, 20] and [21] can closely cooperate with the human, holding an object, lifting a weight, positioning a work piece, etc.. However, finding a trade-off between wearability, efficiency and usability of those bulky extra-arms represents still a challenge. In our opinion, the integration with the human body should be of primary importance in developing wearable devices. For that reason, we decide to investigate how to augment the capabilities of the human hand, instead of developing additional arms.

To the best of my knowledge, the two main independent contributions on the development of extra fingers are from a group at MIT and University of Siena (in the form of this book). The main difference is that they focused on human hand augmentation to increase the workspace of the healthy hand [22] while this book investigated not only augmenting healthy hand [23] but even more on subjects with impairments to recover manipulation abilities [2]. The MIT group presented in [22],

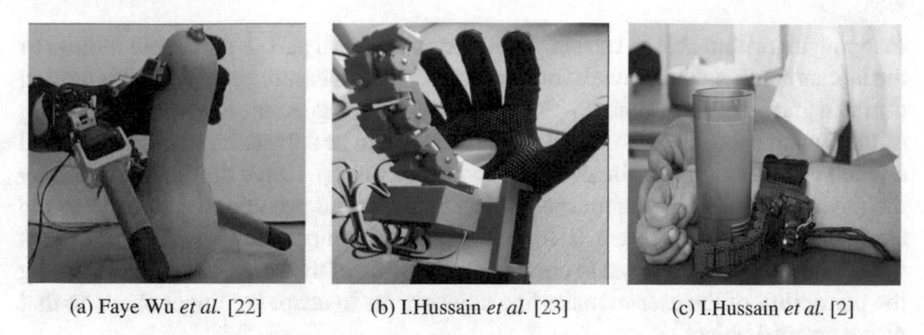

(a) Faye Wu *et al.* [22] (b) I.Hussain *et al.* [23] (c) I.Hussain *et al.* [2]

**Fig. 1.2** Supernumerary robotic fingers developed by MIT and University of Siena to augment and compensate the manipulation abilities of human hand

the design of two supernumerary robotic fingers to augment the abilities of human healthy hand. Supernumerary robotic fingers are mounted on the human wrist to form a 7-fingered hand. 5 human fingers and 2 robotic fingers as shown in Fig. 1.2a. A method for controlling extra robotic fingers in coordination with human fingers to grasp diverse objects has been further developed in [24]. The authors presented a control algorithm, named Bio-Artificial Synergies, enabling a human hand augmented with two robotic fingers, to share the task load together and adapt to diverse task conditions. Postural synergies were found for a seven-fingered hand comprised of two robotic fingers and five human fingers through the analysis of measured data from grasping experiments. However, this approach in the evaluation of synergies for a hybrid human-robot system can be applied only a posteriori, once the system is given, and its application in the synthesis and optimization of robotic devices is not trivial. Although two independent extra fingers showed potentialities in augmenting human hand functions in healthy subjects but size and proposed control strategies limit their possible application as assistive device.

Differently from their approach, this book explored the potential of the supernumerary robotic fingers not only in enhancing the human healthy hand manipulation abilities but more than that, in recovering the abilities of impaired upper limb in case of stroke patients. To this aim, I developed a family of supernumerary fingers ranging from fully actuated to underactuated and soft with their suitable sensoribmotor interfaces. Firstly, we investigated the potentials of extra-finger in healthy subjects. Such devices could give humans the possibility to manipulate objects in a more efficient way, enhancing our hand grasping dexterity/ability. An example of a device devloped with in this book for augmenting human hand is shown in Fig. 1.2b while more details are presented in Chap. 2. Different than the other approaches, we focused on modularity, wearability and portability. To further improve, the wearability the extra robotic finger can be shaped into bracelet when being not used.

The first prototype (shown in Fig. 1.2b) has been presented in [25] together with several examples of the extra-finger applications. Apart from the design issues related to portability and wearability of the devices, another critical aspect was integrating the motion of the extra-fingers with that of the human hand. In [23], we presented a mapping algorithm able to transfer to the extra-fingers a part or the whole motion of the human hand. A commercial dataglove was used to measure the hand configuration during a grasping task. Although this control approach guarantees a reliable tracking of the human hand, there was two main drawbacks to be solved. First, the user lacked a feedback of the robotic finger status and could only perceive the force exerted by the device mediated by the grasped object. The second problem was related to the approaching phase of the grasp. In fact, the algorithm presented in [23], considers the motion of the whole hand to compute the motion of the extra finger, thus limiting the possibility of the user to make fine adjustments to adapt the finger shape to that of the grasped object.

In [26] we addressed these issues by introducing sensorimotor interfaces that can be worn as a ring. The human user receives information through the sensorimotor interfaces about the robotic finger status in terms of contact/no contact with the grasped object and in terms of force exerted by the device. For fully actuated robotic finger, during the grasp approaching phase, we introduced a new control strategy that enables the finger to autonomously adapt to the shape of the grasped object. The more details on sensorimotor interfaces are presented in Chap. 5. The experience gained with healthy subjects was fundamental for the development of the devices for recovering the manipulation abilities of stroke patients. The devices designed for clinical applications must meet specific human factors related to patient conditions and performance criteria. The general guidelines which can be found in the literature include safety, durability, energy efficiency, low-encumbrance, ease of use, error tolerance and configurability. The specific performance criteria and human ergonomics strongly depends on the actual patient conditions and needs. Several experiments with the patients conducted in cooperation with a rehabilitation team led us to outline the main functional requirements and human factors which are considered in the design and development of devices proposed in this book. More details on the ergonomics and functional requirements and how these are met during the design and development of the proposed devices are explained in Chap. 4. One of the underactuated soft robotic fingers designed for recovering the grasping abilities of stroke patient is shown in Fig. 1.2c. Note that, in case of soft fingers, the shape adaptability is achieved through the intrinsic compliance of the flexible material and underactuation. The use of soft supernumerary robotic fingers with stroke patients unfolded and explored their true potential as a new generation of wearable assistive devices which is better justified in the remaining part of this chapter.

Stroke is one of the leading cause to a long term impairment. On average, every 40 s, someone in the United States has a stroke [27]. Impairment of the hand grasping function is one of the common deficits after a stroke: approximately 60% of the stroke survivors suffer from some form of sensorimotor impairment associated with their hand [28]. Different motor impairments can affect the hand both at motor execution and motor planning/learning level. Deficits in motor execution include weakness of wrist/finger extensors, increased wrist/finger flexors tone and spasticity, co-contraction, impaired finger independence, poor coordination between grip and load forces, inefficient scaling of grip force and peak aperture, and delayed preparation, initiation, and termination of object grip [29, 30].

Recovering hand functions is of primary importance during the rehabilitation phase. Many wearable devices have been proposed in the last decade, especially for hand rehabilitation and function recover. A review on robot-assisted approaches to motor neurorehabilitation can be found in [31]. However, most of the devices are designed to increase the functional recovery in the first months of the rehabilitation therapy, when biological restoring and reorganization of the central nervous system can take place. However, only 5–20% of patients show a complete recover of upper limb six months after the stroke [32]. When in the paretic upper limb the motor deficit is stabilized, the rehabilitation consists mainly in ergotherapy, which aims primarily in teaching compensatory strategies that often take advantage of dedicated aids. These strategies may sometimes be neither ergonomic nor ecological, or may even increase pathological motor patterns, usually by worsening tonic flexion at the forearm of the paretic limb [33]. Existing compensatory robotic devices like prostheses, cannot be used for this purpose since the hand of the patient, although frequently with limited mobility, is still present. Early results on the replacement of impaired hand with robotic devices are reported in [34]. However, this potential solution could be much less effective in chronic stroke patients where the whole arm often present a limited residual mobility. Rigid exoskeletons do not accommodate variations in patient skeletal structure or joint misalignment and can produce compression forces on the soft tissue and joints during long-term use [35]. Moreover, most of the proposed exoskeletons result to be quite cumbersome limiting the wearability and portability of the device. Besides exoskeletons and prosthesis and their working principles, it is interesting to study other robotic solutions which can compensate the missing grasping function.

The major part of this book focuses on how supernumerary robotic fingers can be used by stroke patients to improve, recover their missing abilities and can support them in several Activities of Daily Living (ADL) specially in bi-manual tasks.

The majority of daily chores can be classified as "hold-and-manipulate" . They generally require two separate actions: maintaining an object in position and altering some characteristics of the object (e.g. opening a jar). Due to the added advantage of opposable motion, two hands are usually needed to perform such tasks. Those with only one functional hand can modify their behavior and conduct the manipulation with the remaining healthy limb; however, an additional tool is often needed to help secure the object as it is being manipulated.

The main target of the dedicated tools is to let typical bimanual tasks be executed using only the unaffected hand, increasing the functional disparity between the two upper limbs. Moreover, these tools can be difficult to be carried outside structured environments, so limiting their use to rehabilitation facilities or to patient's house.

A possible solution is the use of supernumerary robotic fingers. The aim is to come up with a robotic device that can work together with the paretic upper limb instead of replacing it and without causing any unnatural forces. In robotics, one of the simplest structure that allows to grasp is the gripper. Industrial grippers usually have two fingers and only one degree of freedom. Further simplifying, one finger can be seen as a fixed palm and the other one as an active finger able to restrain the motion of an object. If we consider the paretic upper limb of the patient as a potential fixed palm, what it is really missed is an active finger able to perform the grasp as shown in Fig. 1.2c. This approach recover the ability to grasp and stabilize objects, while keep motivating the patients to use residual mobility of their paretic upper limb. The wearable robotic device can easily be carried by patients even outdoors. Finally, a single robotic device can replace many commercially available tools to perform ADL, since these tools are generally designed to perform a single task. In this view, robotic extra fingers can represent the minimal complexity solution that also guarantee extreme wearability and that do not require to be coupled with human impaired limbs in order to compensate for missing capabilities.

The supernumerary finger can be used together with the paretic hand/arm, to constrain the motion of the object. The device can be worn on the user's forearm by means of an elastic band. The systems acts like a two-finger gripper, where one finger is represented by the robotic finger, while the other by the patient's paretic limb as shown in Fig. 1.1c. The preliminary results with patients are presented in [36]. The patient can regulate the finger flexion/extension through a wearable switch embedded in a ring worn on the healthy hand. Two possible predefined motions can be chosen to obtain either a precision or a power grasp. In addition to the switch, the proposed ring interface also embeds a vibrotactile motor able to provide the patient with information about the force exerted by the device. Although, the ring based control approach resulted simple and intuitive. However, involving the healthy hand in the control of the extra-finger can interfere when performing bimanual tasks or when grasping an object, since the patient has to be careful not to activate the switch unintentionally.

In [37] we introduced an EMG interface using frontalis muscles to control the finger flexion/extension motion while maintaining the principle of simplicity of the switch. This approach leaves the patient free to use his or her healthy hand. Moreover, the frontalis muscle is always spared in case of a motor stroke either of the left or of the right hemisphere due to its bilateral cortical representation. The user can contract this muscle by moving the eyebrows upwards. The design of earlier presented robotic finger was fullyactuated and realized through the rigid structure. The resulting size and weight affect the device portability, wearability and robustness. For this reason, we designed and developed soft underactuated cable driven robotic fingers to guarantee high wearability and portability. In [2], we presented the Soft-SixthFinger which is passively compliant thanks to its flexible joints. Only one motor is used to regulate the device flexion through a tendon-driven actuation. The advantage of this solution with respect to the previous versions of the device are threefold. The compliance in the structure simplifies the control of the device that can passively adapt to the shape of the grasped object [38]. Flexible joints increase also the robustness against undesired contacts with the environment [39]. Finally, the tendon-driven underactuation reduces the total weight of the device increasing its portability and wearability. To further improve the wearability and portability, the device can be worn as a bracelet when it is not used. The patient can switch from the bracelet to the operative position by using his or her healthy hand. Moreover, in the previous versions of the device, the batteries and all the electronic boards for the control were worn in the paretic arm by means of an elastic band. The arising weight limited the mobility on the patient's paretic arm. In the Soft-SixthFinger, the battery bank and the electronics board were moved from the paretic arm to a box that can be fixed in the user belt. The result is a sensible reduction of the weight the user has to support with the paretic arm. The soft robotic fingers are the results of experimental sessions with chronic stroke patients and consultations with clinical experts. More details on the design, analysis, fabrication, experimental characterization and evaluation are illustrated in Chap. 4. In a recent work, presented in Chap. 7, the combination of the soft supernumerary robotic finger with the zero gravity arm support, the SaeboMAS(Saebo, Charlotte, USA) is proposed. The proposed system can be used during the rehabilitation phase when the arm is potentially able to recover its functionality, but the hand is still not able to perform a grasp due to the lack of an efficient thumb opposition. The overall system also act as a motivation tool for the patients to perform task-oriented rehabilitation activities. The patient can closely simulate the desired motion with the non-functional arm for rehabilitation purposes, while having a grasp compensation with the help of the supernumerary robotic finger.

Another important question that we are investigating is the neural embodiment of supernumerary robotic fingers i.e. how the human brain potentially perceives them. The neural embodiment represents a hierarchically higher level interaction of human with the robotic fingers. It considers whether the extra robotic devices could enter into the individuals body schema. The first evidence that the perceptual illusion of owning an artificial hand could be induced, was provided by the rubber hand illusion in 1998 [40]. In 2011, Guterstam [41] demonstrated the possibility of inducing the perceptual illusion of having a supernumerary right hand. Their findings suggest that the neural embodiment of a supernumerary limb is possible if it is aligned with the body in an anatomically similar fashion as the real limbs. In addition, it has been demonstrated that the size of the incorporated body part is not important and ownership illusion can be induced towards very small or large bodies [42]. Other studies show that although multiple supernumerary limbs can be incorporated into the bodily image (i.e. the sense of ownership towards the supernumerary limb), only one can be included in the body schema (i.e. the ability to control the supernumerary limb) [43, 44].

Previous fMRI studies showed that the human Broca's area plays a fundamental role in processing not only language, but even tool use. This suggests that language and tool use share computational principles for processing complex hierarchical structures common to these two abilities [45]. Developing a suitable strategy for controlling an extra limb, whether it is a finger or an arm, is of high interest for various fields: for example, in [46] authors studied how a surgeon can use his foot to control an extra arm, at least in a virtual environment.

Since it is important that the operator keeps focused on the task he has to perform, the control paradigm of the robot should be felt natural to him, as if the supernumerary limb was part of his body. This is why the study of embodiment is a fundamental pillar of the research of supernumerary robotic limbs.

# Chapter 2
# Enhancing Human Hand Manipulation Abilities Through Supernumerary Robotic Fingers

*I see technology as being an extension of the human body*

*David Cronenberg*

One of the new targets of wearable robots is not to enhance the lift strength far above human capability by wearing a bulky robot, but to support human capability within its range by wearing lightweight and compact robots. A new approach regarding robotic extra-fingers is presented in this chapter. In particular, different design guidelines to realize robotic extra-fingers for human grasping enhancement are proposed. Such guidelines were followed for the realization of three prototypes obtained using rapid prototyping techniques, i.e., a 3D printer and an open hardware development platform. Both fully actuated and underactuated solutions have been explored. In the proposed wearable design, the robotic extra-finger can be worn as a bracelet in its rest position. The availability of a supplementary finger in the human hand allows to enlarge its workspace, improving grasping and manipulation capabilities. To control the motion of extra fingers, an object-based mapping algorithm is proposed by interpreting the whole or a part of the hand motion in grasping and manipulation tasks.

## 2.1 Introduction

Wearable robots are expected to work closely, to interact and collaborate with people in an unstructured environment [8]. By definition, a wearable robot is a mechatronic system designed around the shape and function of the human body, with segments



I. Hussain and D. Prattichizzo, *Augmenting Human Manipulation Abilities with Supernumerary Robotic Limbs*, Biosystems & Biorobotics 26,
https://doi.org/10.1007/978-3-030-52002-1_2

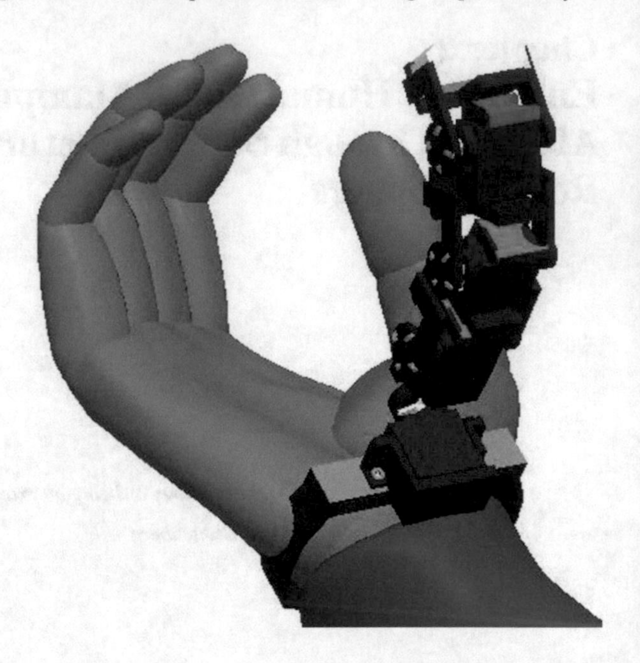

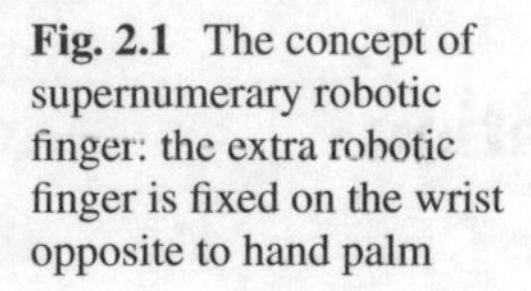

**Fig. 2.1** The concept of supernumerary robotic finger: the extra robotic finger is fixed on the wrist opposite to hand palm

and joints corresponding to those of the person it is externally coupled with [9]. Such definition was coined for exoskeletons, whose main purposes were enhancing human body force and precision capabilities [47] or helping in rehabilitation processes [48]. The progress in miniaturization and efficiency of the mechanical and sensing components has extended the field of wearable robotics to new devices which can be seen as extra limbs. In [21] for instance, two additional robotic arms worn through a backpack-like harness are presented. However, finding a trade-off between wearability, efficiency and usability of those bulky extra-arms represents still an issue.

We started to investigate how to enhance the capability of the human hand by means of wearable robots [49]. The goal was to integrate the human hand with additional robotic fingers as represented in Fig. 2.1 for the case of a sixth finger.

Adding wearable robotic fingers could give humans the possibility to manipulate objects in a more efficient way, enhancing our hand grasping dexterity/ability. We designed and developed both fully actuated and underactuated solutions for the robotic extra finger. The corresponding prototypes can be realized by using standard rapid prototyping techniques (see Fig. 2.2). Both the solutions share the idea of wearability. The devices can be worn as bracelet when they are not activated, and can pop up when needed.

Together with the design issues related to portability and wearability of the devices, another critical aspect is integrating the motion of the extra-fingers with the human hand [23, 25, 50]. In fact, demonstration-based algorithms as that presented in [21] fail in generality and adaptability to new tasks, while classical techniques for

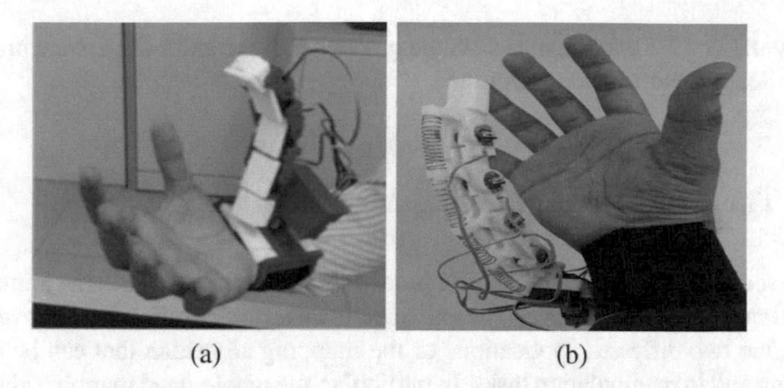


**Fig. 2.2** Fully actuated and underactuated robotic extra fingers prototype. The extra fingers interacts and coordinate with the human fingers in grasping and manipulation tasks

exoskeleton controllers, based on the reading of some bio-signal like EMG, limit the wearability [51].

This chapter also presents a mapping algorithm able to transfer to the extra-fingers a part or the whole motion of the human hand. The algorithm extends the method proposed in [52] to the case of a human hand augmented with robotic extra-fingers. The mapping algorithm is based on the definition of a set of reference points on the so called *augmented hand* which includes the human hand and the robotic extra-fingers. A virtual sphere is defined as a function of the reference points. When the human hand fingers are moved, the virtual sphere is moved and deformed. The robotic extra-fingers are then actuated so that the reference points on them follow the virtual sphere transformation. The mapping algorithm can be applied considering the whole human hand in the definition and in the transformation of the virtual sphere, but it is possible to adopt only a part of it, for example three fingers. In this case the remaining fingers, not involved in the mapping process, can be used to perform another task. As an illustrative example, let us consider the task of holding and opening a bottle. This task can be difficultly performed with only one hand. If a robotic extra finger is available, it can be used together with the middle, ring and little finger to hold the bottle, while the index and the thumb can unscrew the cap.

As a case study, we developed a modular finger that can be worn on the wrist with the help a rubber band. The structure of the modules is obtained using rapid prototyping techniques, while the active degrees of freedom (DoFs) are realized with servomotors. The main aim of the prototype described in this work is to make human hand more symmetric, so that when the human fingers close (when phalanges flex), the extra-finger reflects exactly this motion. This extension is going to increase the human hand workspace and its grasping abilities at the same time.

The rest of the chapter is organized as it follows. Section 4.3 illustrates the control framework for a robotic extra-finger. Section 2.3 shows the numerical simulations for performance evaluation of the proposed control algorithm. In Sect. 2.4 the design guidelines for robotic extra-fingers are detailed. Section 2.4.1 describes the fully actuated finger, while in Sect. 2.4.2 the underactuated version of the finger is shown.

Finally in Sect. 2.5 experiments with the prototype of the extra-fingers are presented and in Sect. 2.6 conclusion is outlined.

## 2.2 The Task-Based Mapping Algorithm

In this section we describe the general procedure proposed to control the joints of the extra-fingers. The algorithm can be extended to an arbitrary number of extra-fingers. We define two different applications of the mapping algorithm that can be used in grasping and in manipulation tasks. In particular, the whole-hand mapping algorithm considers all the human hand fingers to compute the motion of the robotic devices. This algorithm is particularity suitable for grasping task especially in the approaching phase. The other application is the partial-hand mapping algorithm which considers only some of the human fingers to compute the motion of the robotic fingers. This last approach is useful in manipulation task when a part of the hand collaborate with the robotic device to hold an object, while the remaining fingers perform a different operation, e.g. unscrew a cap while holding a bottle.

### *2.2.1 The Whole-Hand Mapping Algorithm*

Let us define a reference frame $S_0$ on the hand. Its origin, $O$ is in the wrist center of rotation, the $z$ axis is perpendicular to hand palm plane, the $x$ axis is the intersection between the sagittal and the transverse plane, pointing towards the little finger, the $y$ axis is consequently defined [53]. Let us indicate with $p_i^h \in \Re^3$, $i = 1, \ldots, n_h$ the coordinates of reference points on the reference human hand, expressed w.r.t. $S_0$. In this paper we choose as reference points the five fingertips of the human hand, therefore $n_h = 5$. Note that, the mapping algorithm proposed in [52] does not constraint the choice of the reference point positions. However, the fingertips are a preferable selection since they are at the end of the kinematic chains represented by the fingers and, thus, their positions contain information about all the joint values.

Let us assume that all the $p_i^h$ can be measured or evaluated, for instance by means of an instrumented glove and direct kinematic computation.

Let define as the *augmented hand* the system composed by the hand with its five fingers and those artificial. On all the robotic fingers, a reference point is placed at the tip. The augmented hand is then characterized by a set of reference points that is the union of the set of reference points on the human fingers and the set of points on the artificial ones. The reference points for the augmented hand are then $p_i^a \in \Re^3$, $i = 1, \ldots, n_a$, with $n_a > n_h$. The first $n_h$ reference points are the same of the human hand, i.e. $p_i^h = p_i^a \ \forall i \leq n_h$ while the remaining $n_a - n_h$ points are those relative to the extra-fingers. Let us indicate with $\mathcal{O}$ the minimum volume bounding sphere containing all the $n_a$ reference points of the augmented hand. Let $o^h$ indicate its center and let $r^h$ be its radius. Let us define a reference frame $S_1$ on the virtual sphere, whose origin is in the sphere center and whose axis are, in the reference starting position, parallel to $S_0$ axis. The reference starting position is arbitrary and do not affect the mapping procedure as detailed in [52].

Assume the augmented hand in its starting position at time instant $t = t_0$. Assume also that at time instant $t = t_0 + \delta t$ the reference point coordinates $p_i^h$ change due to the motion of the human hand. Let us indicate with $\Delta p_i^h \in \Re^3, i = 1, \ldots, n_h$ a vector containing such coordinate variations. This displacement produces a transformation in the virtual sphere, that in this paper we approximate as the combination of a rigid body motion and an isotropic deformation. The rigid body motion can be furthermore represented as the combination of a translation $\Delta o^h \in \Re^3$ and a rotation $\Delta \Phi \in \Re^3$. The rotation term $\Delta \Phi \in \Re^3$ is defined as

$$\Delta \Phi = [\Delta \phi, \Delta \theta, \Delta \psi]^{\mathrm{T}},$$

where $\Delta \psi$ represents the rotation w.r.t. $x$ axis, $\Delta \theta$ represents the rotation w.r.t. $y$ and $\Delta \psi$ represents the rotation w.r.t. $z$ axis. It is worth to recall that, even though the rotations between reference frames are not commutative, and the rotation order is therefore important, if the rotation angles are *small*, the rotation order is not significant. The non rigid isotropic deformation can be described by the parameter $\Delta s \in \Re$ defined as

$$\Delta s = \frac{\Delta r}{r}.$$

With these assumptions, we can express the displacement of each reference point on the human hand as follows

$$\Delta p_i^h = \Delta o^h + \Delta \Phi \times \left(p_i^h - o\right) + \Delta s \left(p_i^h - o\right). \tag{2.1}$$

In this paper we do not consider other types of transformations for the sake of simplicity, however the method can be integrated to include a non-isotropic transformation, as described in [54], and also shear deformations [55].

Equation (2.1) can be applied to all the reference points $p_i^h$, leading to the following linear system

$$\Delta p^h = A \Delta \xi, \tag{2.2}$$

where $\Delta p_h = \left[\Delta p_1^{h\mathrm{T}}, \ldots, \Delta p_{n_h}^{h\mathrm{T}}\right]^{\mathrm{T}} \in \Re^{3n_h}$ is a vector collecting all the reference point displacements, $\Delta \xi = \left[\Delta o^{\mathrm{T}}, \Delta \Phi^{\mathrm{T}}, \Delta s\right]^{\mathrm{T}}$ is a $7 \times 1$ vector containing the unknown parameters describing the sphere transformation, including the translation term $\Delta o^h$, the rotation term $\Delta \Psi$ and isotropic deformation term $\Delta s$, finally, $A \in \Re^{3n_h \times 7}$ is the linear system matrix, defined as

$$A = \begin{bmatrix} A_1 \\ \cdots \\ A_{n_h} \end{bmatrix},$$

in which each submatrix $A_i \in \Re^{3\times 7}$ is defined as[1]

$$A_i = \left[ I \; -s(p_i^h - o) \; (p_i^h - o) \right].$$

The linear system in Eq. (2.2) can be solved, to find

$$\Delta\xi = A^{+}\Delta p^h + N_A\psi \tag{2.3}$$

where $A^{+}$ denotes a generic pseudo-inverse of $A$ matrix, while $N_A \in \Re^{7\times h}$ represents a basis of $A$ matrix nullspace, whose dimension is $h \geq 0$, and $\psi$ is an arbitrary $h$-dimensional vector parametrizing the homogeneous solution of the system. When $h > 0$, the vector $\psi$ can be defined to optimize a cost function that can be defined on the basis of the task, e.g. when a grasping task is performed, we would need to assure grasp stability, maximizing $\Delta s$ magnitude, while in object manipulation tasks, in which the contact forces should be constant, we should maximize $\Delta o$ and $\Delta\Phi$ and minimize $\Delta s$. Once the sphere transformation parameters have been evaluated we need to generate the command signals for the extra-finger joints. What we impose is that also the reference points of the extra-fingers move according to the transformation parameters computed on the virtual sphere. In particular, we consider that for all $p_i^a$ with $n_h < i \leq n_a$,

$$\Delta p_i^a = \Delta o^h + \Delta\Phi \times \left(p_i^a - o\right) + \Delta s \left(p_i^a - o\right), \tag{2.4}$$

where the parameter $\Delta o$, $\Delta\Phi$ and $\Delta s$ are those computed in Eq. (2.3). The joints of the robotic fingers are then controlled to obtain on the fingertips the displacement evaluated in Eq. (2.4). Let $n_q$ indicate the number of joints of the artificial fingers and let $J_a \in \Re^{3(n_a - n_h)\times n_q}$ indicate the finger Jacobian matrix, we can evaluate finger joint displacements as

$$\Delta q_a = J_a^{+}\Delta p_i^a + N_J\chi, \tag{2.5}$$

where $J_a^{+}$ denotes a generic pseudoinverse of robotic finger Jacobian matrix, $N_J \in \Re^{n_q\times n_\chi}$ is a matrix whose columns form a basis of $J_a$, $n_\chi$ represents the dimension of robotic finger redundancy space, $n_\chi \geq 0$, and $\chi \in \Re^{n_\chi}$ is a $n_\chi$-dimensional parameter parametrizing the homogeneous part of the solution, which represents the redundant motions of the robotics fingers. Due to the computational time and the dynamics of the robotic fingers, the joint displacements $\Delta q_a$ computed at the time sample $t = t_0$ is effectively executed at time $t = t_0 + \delta t$, where $\delta t$ indicates the sampling time. This delay does not affect the mapping procedure. The mapping algorithm is

---

[1]For any three-dimensional vector $v = [v_1,\; v_2,\; v_3]^{\mathrm{T}}$, $s(v)$ indicates the skew matrix associated with vector $v$, i.e.

$$s(v) = \begin{bmatrix} 0 & -v_3 & v_2 \\ v_3 & 0 & -v_1 \\ -v_2 & v_1 & 0 \end{bmatrix}.$$

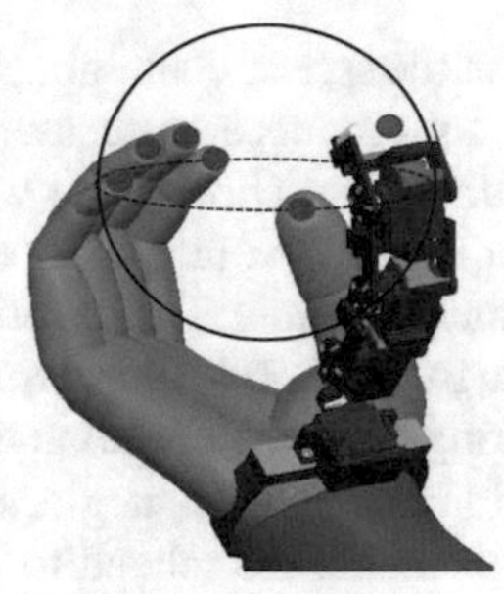

(a) The fingertips of the augmented hand are selected as reference points. The virtual sphere is defined as the minimum volume sphere containing all the reference points, $t = t_0$.

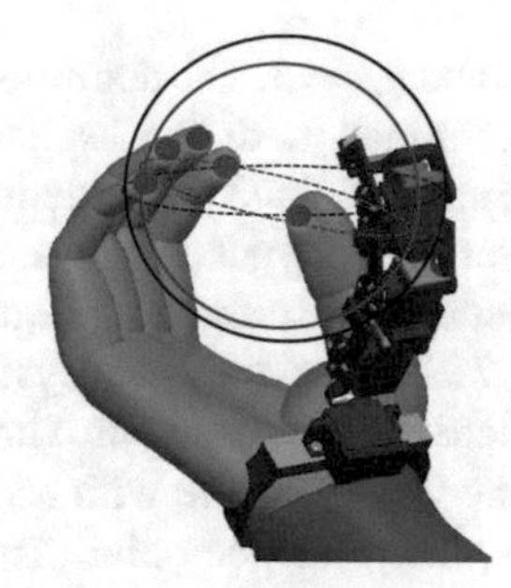

(b) The motion of the human hand displaces the reference points and thus deforms and moves the virtual sphere. The reference angle for the joints of the extra–fingers are computed, $t = t_0$.

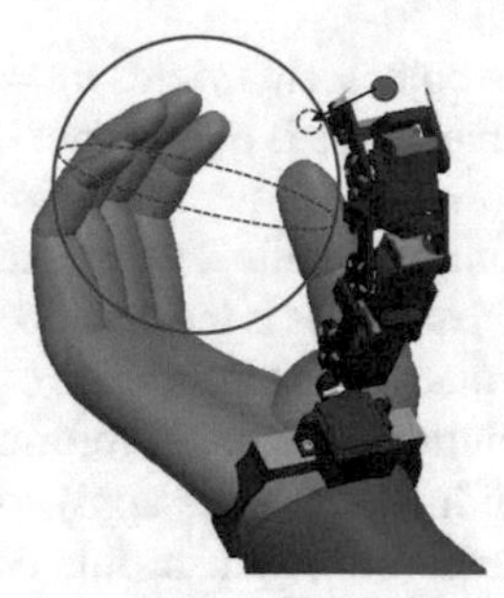

(c) The extra-finger is moved according to the rigid body motion and the deformation of the virtual sphere, $t = t_0 + \delta t$.

**Fig. 2.3** The mapping algorithm

pictorially represented in Fig. 2.3. Once the robotic finger has been actuated, at the time $t = t_0 + \delta t$ the virtual sphere is re-calculated and the procedure is repeated.

### 2.2.2 *The Partial-Hand Mapping Algorithms*

The human hand can perform complex tasks that involve both grasping and manipulation. As an example, let us consider the task of open a bottle. Depending on the size of the bottle and on user ability, either the task can be performed using only one hand (the bottle is held by the middle, ring and little fingers and the hand palm, while the thumb and the index are in charge of opening it), or the task cannot be performed using only one hand, since the three fingers and the palm are not sufficient to guarantee grasp closure properties [56]. The extra-finger could be used in this example to let the hand hold the bottle, while the thumb and the index unscrew the cap. In such scenario, it would be useful to relate the motion of the robotic fingers only to some of the hand fingers, for instance the middle, ring and pinkie fingers. This solution leaves the thumb and the index able to act independently and does not affect grasp tightness.

Framing in the procedure previously described, we have $n_h = 3$ and $n_a = 4$. It is still possible to define a sphere passing through the reference points, and $A$ matrix dimensions are $9 \times 7$, so the linear system defined in Eq. (2.2) can be solved to obtain the transformation parameters $\Delta o$, $\Delta \Phi$ and $\Delta s$, that can be used to control the motion of the robotic finger. From the theoretical point of view, it is then possible to control the robotic extra-finger mapping only three of the five human fingers.

More complex manipulation tasks could require the use of three fingers, namely the thumb, index and middle, leaving the remaining two and the robotic one to hold the object, This yields $n_h = 2$ and $n_a = 3$. The definition of the sphere containing the three points is a priori not determined (to define a sphere, four points are necessary). However, among the infinite spheres that can be defined, we can choose that with minimum radius, whose center is in the geometric center of mass of the points and whose radius is the distance between the reference points and the center. The resulting $A$ matrix has dimensions $6 \times 7$ and, thus, the linear system in Eq. (2.2) admits infinite solutions. We could reformulate the problem approximating the transformation that the human fingers apply to the virtual sphere with a rigid body motion, neglecting the deformation. In this case, the reference point displacements are related to the sphere transformation parameters by

$$\Delta p_i^h = \Delta o + \Delta\Phi \times (p_i^h - o),$$

for $i = 1, 2$, that can be rewritten as

$$\Delta p^h = A_{rb} \Delta\xi_{rb},$$

in which $\Delta p^h = [\Delta p_1^h,\ \Delta p_2^h]^{\mathrm{T}}$, $\xi_{rb} = [\Delta o,\ \Delta\Phi]^{\mathrm{T}}$ and

$$A_{rb} = \begin{bmatrix} I & -s(p_1^h - o) \\ I & -s(p_2^h - o) \end{bmatrix} \in \Re^{6\times 6}.$$

Matrix $A$ is invertible, unless the reference points are coincident, and then we can estimate the sphere transformation parameters as

$$\xi_{rb} = A^{-1} \Delta p^h.$$

Such parameter are used to project the sphere transformation onto the robotic extra-fingers, as previously detailed. It is worth to observe that $A_{rb}$ matrix corresponds to the transpose of the classical Grasp matrix, [57]. We furthermore observe that, if the augmented hand is grasping an object with two human and a robotic fingers, we impose to the robotic finger to follow the human ones without changing the internal contact forces.

## 2.3 Numerical Simulation

The performance of the proposed control algorithm has been evaluated through a series of numerical simulations using *SynGrasp*, a Matlab Toolbox that aims at simulating and analyze hand grasping. The toolbox includes functions to define the hand kinematic structure and the contact points with a grasped object. New functions have been added to describe a hand model where also extra-fingers are considered.

Concerning the human hand, in this work we consider the 20 DoFs hand kinematic model detailed in [58, 59].

The structure of the human hand has been augmented with the robotic extra-finger prototype model, which has 4 DoFs: one for the finger abduction/abduction motion and three simulating the phalanges' flexion/extension motion. The resulting augmented hand structure has then 24 DoFs. To control the robotic extra-finger, the mapping algorithm previously illustrated has been implemented. The numerical simulations were devoted to evaluate the role of the robotic extra-finger in grasping tasks. We considered the grasping of simple shape objects, namely spheres, cubes, and cylindrical pipes, with different sizes, varying from 40 to 140 mm the radius/edge. We analyzed and compared three configurations: (1) the human hand, (2) the complete augmented hand, (3) the augmented hand in which the robotic finger is controlled with only three fingers of the human hand, namely the middle, ring and pinkie fingers.

For each grasp, we analyzed the values of one of the grasp quality indexes, in this case the *grasp isotropy index* ($GII$) discussed in [60], defined as the ratio between the minimum $\sigma_{min,G}$ and maximum $\sigma_{max,G}$ singular value of the grasp matrix $G$,

$$GII = \frac{\sigma_{min,G}}{\sigma_{max,G}}.$$

The index $GII$ measures the contribution of the contact forces to the total wrench exerted on the object, and varies from 0 to 1. Its optimal value, i.e. $i = 1$, corresponds to an isotropic grasp, in which the magnitudes of the internal contact forces are similar. Further details on the definition and evaluation of grasp matrix $G$ are available in [57].

The results, in terms of grasp quality matrix, are summarized in Fig. 2.4, showing the $GII$ index as a function of the object size, for different grasping conditions and different objects. The robotic extra-finger improves the grasp quality index, in particular for large size objects. These results confirm the intuitive observation that an additional robotic thumb improves grasp capabilities of the human hand enlarging its workspace. We furthermore observe that activating the robotic finger using the data from only three fingers of the human hand we obtain grasps with a slightly lower quality measure $GII$ w.r.t. the complete augmented hand, however such values are generally higher than those obtained with the human hand, especially for larger objects. Figure 2.5 shows some examples of grasps realized with the human and the augmented hand.

## 2.4 Design Guidelines for Robotic Extra-Fingers

Wearability is the common target of the design guidelines for robotic extra fingers. In the solutions proposed in this paper, the finger can be easily worn by the user by means of an elastic band. It wraps up on the wrist as a bracelet when it is not used and it pops-up when it is actuated, as shown in Fig. 2.7. A push button is used to switch between the two positions in the presented prototypes. The extra-fingers should be able to mimic the motion of human hand's fingers so in both the designs the kinematics of human fingers are partially resembled. Joints in the finger are connected to partially

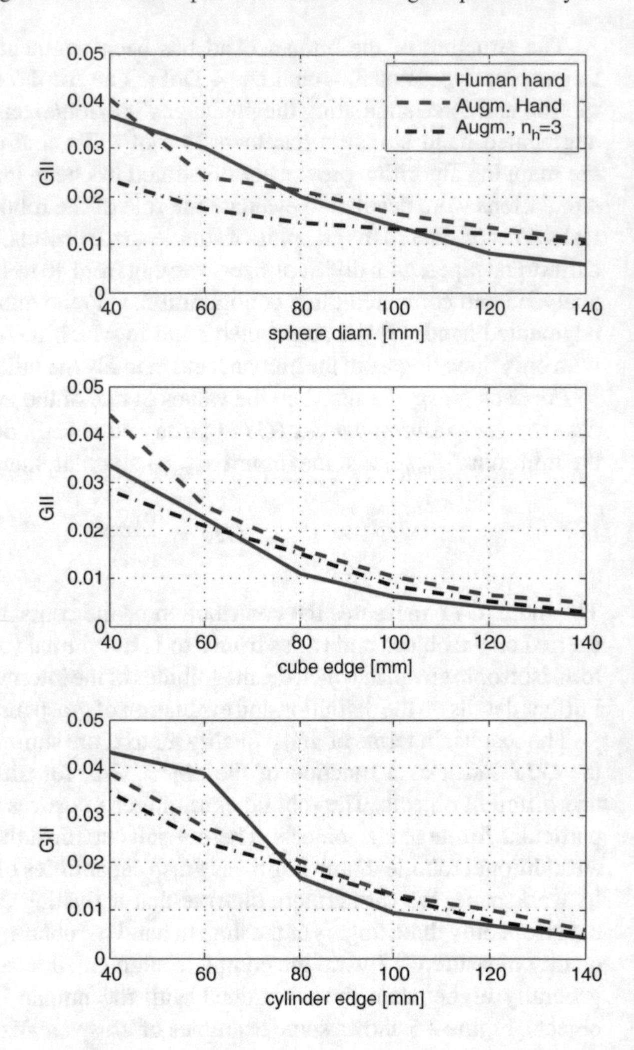


**Fig. 2.4** Grasp isotropy index, i.e. ratio between the minimum and maximum singular value of grasp matrix $G$, obtained for different object shapes and sizes, with the human, the augmented hand, and the augmented hand with $n_h = 3$

resemble the human finger mechanical structure. Our fully actuated finger design is based on this principle while ensuring the structure design to have reasonable size, workspace and use of 3D printed material for rapid prototyping and light in weight. Hand fingers, excluding the thumb, consists of four phalanxes connected by three joints [61, 62]. The structure of the thumb is different, since it has two joints at the base for the anterposition or retroposition combined with the radial or palmar abduction motions.

The other fingers are capable of both adduction-abduction and flexion-extension motions. The proximal and distal interphalangeal articulations can have only flexion/extension motion capabilities and typically are represented with a single DoF revolute joint. The metacarpal joints have both adduction/abduction and flexion/extension motion capabilities and can be modeled as a 2 DoFs joint composed

**Fig. 2.5** Example of grasps evaluated during the numerical simulations. First row: grasping with the human hand; second row: grasping with the augmented hand; third row: grasping with the augmented hand in which the index and the thumb are not active. First column: grasping a sphere; second column: grasping a cube; third column: grasping a cylindrical pipe

of two revolute joints with orthogonal rotation axis (universal joint). We choose this kinematic structure to design the robotic extra finger. In the fully actuated version of the device one motor is adopted to actuate each DoF of the robotic-finger, so two motors at the base realize the adduction/abduction and flexion/extension motions, while one motor for each interphalangeal joint is necessary. The finger's kinematic model is typically approximated by using simple revolute joints. This approximation is an effective means of modeling, as these are in fact the same as compared to proximal and distal joints of humans. Although for metacarpal joints simple revolute joint approximation is not valid, adduction/abduction motion are approximated by adding an additional actuator.

In the rest of the section we will describe in details the two proposed kinematic structures.

## 2.4.1 Fully Actuated Finger

The fully actuated finger has a modular structure. The final size of the device and the number of DoFs can be selected according to dimension of the user's hand. Modularity also offers robustness considering that robot parts are interchangeable. Each unit module consists of a servo motor and two plastic parts. It results in one DoF and its dimensions are $42 \times 33 \times 16$ mm. The same modules can be connected through screws to obtain more complex kinematic chains like the modular hands described in [63]. The cad design of unit module and complete fingers are shown in Fig. 2.6. The position of the extra-finger on the wrist is crucial to obtain the

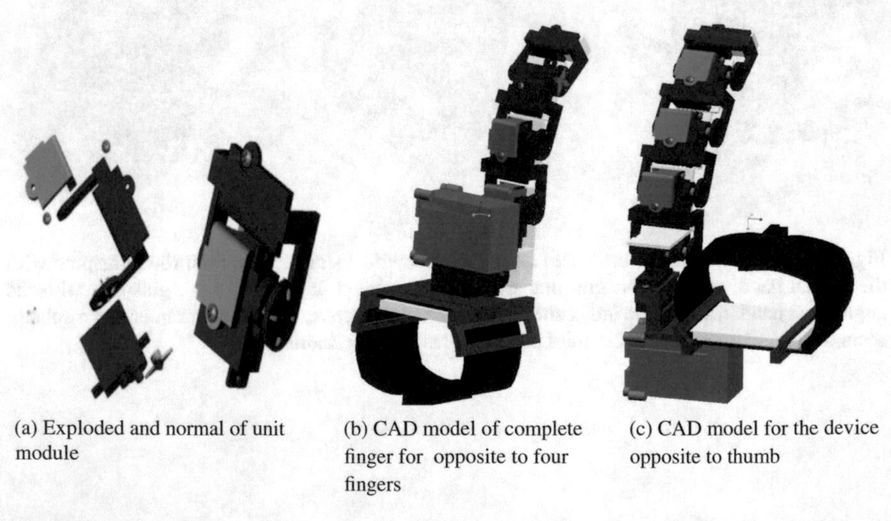

(a) Exploded and normal of unit module

(b) CAD model of complete finger for opposite to four fingers

(c) CAD model for the device opposite to thumb

**Fig. 2.6** CAD Design of unit module and complete fully actuated fingers for both positions are shown. The single DoF modules are connected through a wrist rubber band to wear the device

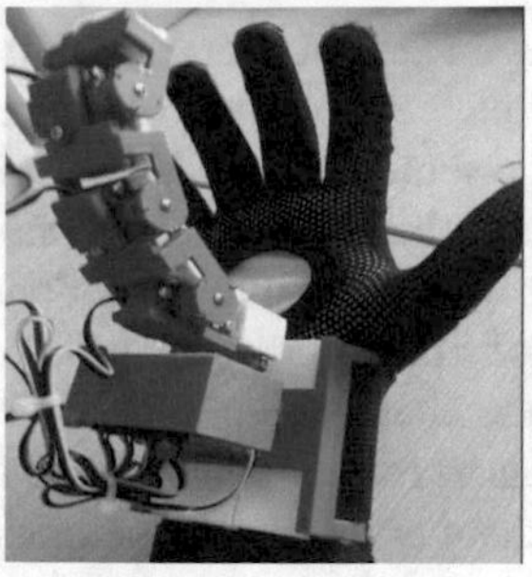
(a) Central configuration.

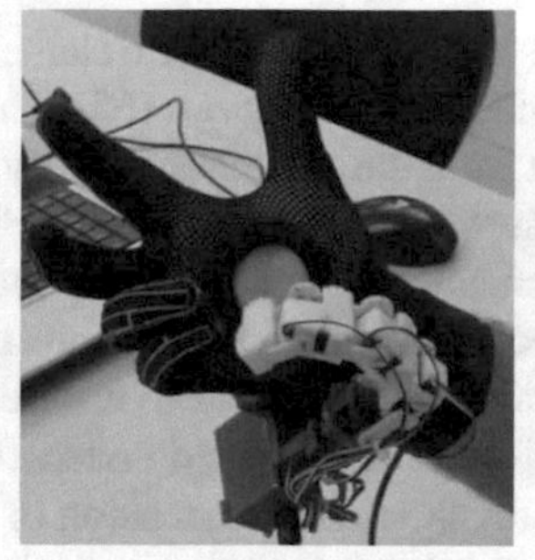
(b) Second thumb configuration.

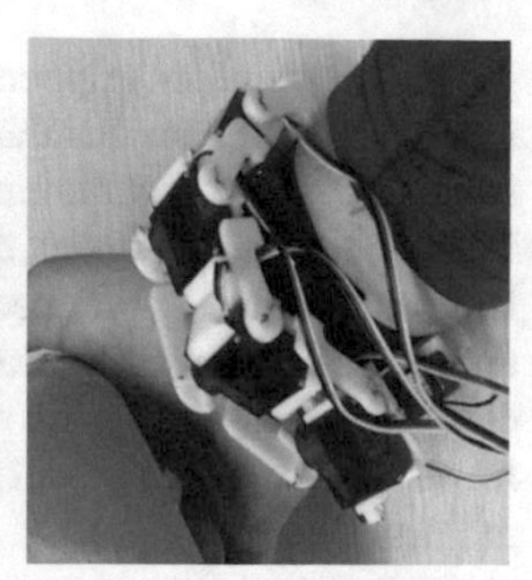
(c) The extra-finger rest position.

**Fig. 2.7** The robotic extra finger in two possible configurations on the wrist

**Table 2.1** Denavit–Hartenberg parameters for the extra-finger, where $a = 42$ mm

| Joint | Id. | $a_j$ | $\alpha_j$ | $d_j$ | $\theta_j$ |
|---|---|---|---|---|---|
| MCP | 1 | 0 | $-\pi/2$ | 0 | $\theta_1$ |
| MCP | 2 | $a$ | 0 | 0 | $\theta_2$ |
| PIP | 3 | $a$ | 0 | 0 | $\theta_3$ |
| DIP | 4 | $a$ | 0 | 0 | $\theta_4$ |

desired performance in different applications. At the moment, we have considered two possible positions on the wrist that we called *central* and *second thumb*, as shown in Fig. 2.7. In the *central* configuration, the finger is placed in the center of the wrist, opposite to the four fingers of human hand so to enlarge the hand workspace. The finger is a four DoFs modular structure. Three DoFs are obtained considering three modules in a pitch-pitch connection, which is used to replicate the flexion/extension motion of the fingers. These modules are connected through a base servo motor to a rubber band that allows to wear the device on the wrist. The base servo motor reproduces the abduction/adduction capability of the fingers.

In the *second thumb* configuration, the aim is to provide an additional thumb to the hand, placed next to the pinkie finger.

The CAD models and the prototypes realized for both the positions are shown in Figs. 2.6 and 2.7. Table 2.1 shows the D-H parameters of the modular device. All the electronics is enclosed in a 3D printed box attached to the finger to make it wearable.

### 2.4.2 *Underactuated Compliant Finger*

For the underactuated finger, we developed only the *second thumb* version, since it resulted to be the more suitable to fit the user wrist. Due to its passively compliant structure, the finger can adapt to the shape of the grasped objects [64]. Underactuated fingers in compliant grippers generally use elastic elements in the not actuated joints, which are then compliant and passively driven [65]. The concept of underactuation in robotic fingers is different from the one that is usually presented in robotic system. In an underactuated finger, the actuation torque $\tau_a$ is typically transmitted to the phalanges through suitable mechanical design, e.g., four-bar linkages, pulleys, tendons and gears etc. Since underactuated fingers have more DoFs than actuators, passive elements are used to constrain the finger and to ensure the shape adaptation of the finger to the grasped object [66]. The proposed underactuated robotic finger has two motors. One motor is used for the flexion/extension of the whole finger. The other motor accounts for adduction/abduction motion and it is also used to switch between rest and operative position. The finger has four phalanges and a base part. All the phalanges are connected by screws. Bearings are used to reduce the friction between phalanges. Also in this case we have pursued modularity and wearability concepts in designing a device that can be worn as a bracelet in its rest position. The base part of the finger consists of a motor, a switch, electronic circuitry and a wrist band. The finger's drive mechanism is a nylon wire, which connects the outermost phalanges with the motor through a pulley. If no torque is applied, the finger is completely straight. When the motor is activated the wire is pulled and the finger bends to grab the object. The springs placed on the back of the finger are used to bring it to its original position when required.

In order to derive the kinematics equation of the under-actuated finger, we have assumed the planar motion of the finger, i.e. abduction/adduction is not considered. We can estimate the position of the finger in the relative plane with the help of forward kinematics, provided that the link lengths and joint angles are known. According to the dimensions and geometric properties of the realized prototype, the total spring stiffness for the four springs is 3.5016 N/cm. The total deflection required to bend the finger completely is 1.4 cm and thus corresponding force is $F = 4.9$ N. The pulley on which the wire is rolled has a radius of 0.7 cm. Thus the motor torque

**Fig. 2.8** CAD models and prototype of the underactuated finger. Passive unit modules can be added/removed to adjust the length of finger

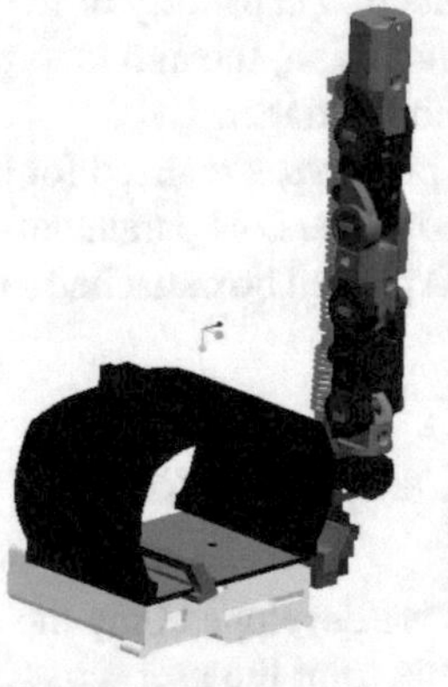

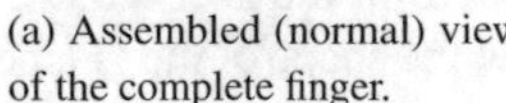

(a) Assembled (normal) view of the complete finger.

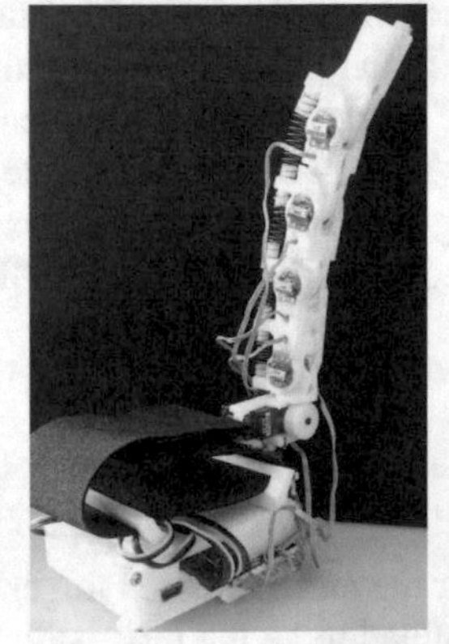

(b) Prototype of the underactuated robotic finger.

required to fully flex the finger is 0.35 kg cm. The motor used is HS-55 micro lite servo which has operating torque of 1.1–1.3 kg cm. The CAD and real prototype of the underactuated compliant finger are reported in Fig. 2.8.

## 2.5 Experiments

We performed several experiments to test the capability of the developed fingers to enlarge hand workspace in grasping tasks and increase human hand dexterity We focused mainly into two different tasks that we called: *anatomically impossible* grasps and *ulnar* grasps. In the former case, we tried to grasp objects which cannot be grasped using only one hand. In the latter, we tried to grasp objects only using the ring and the pinkie fingers opposite to the sixth finger. In this case the upper part of the hand (thumb, index and medium fingers) is left free to do another task allowing, for instance, to hold multiple object in one hand, to unscrew a bottle cap with a single hand, etc.

(a) Grasping multiple objects in one hand: fully actuated finger in second thumb configuration.

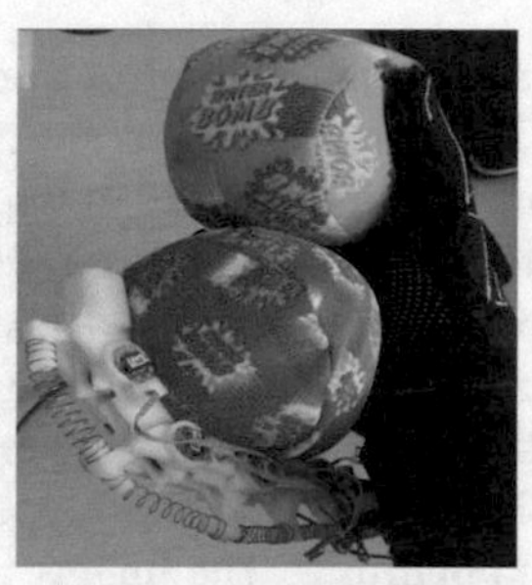

(b) Grasping multiple objects in one hand: underactuated finger.

(c) Grasping box: fully actuated finger in second thumb configuration.

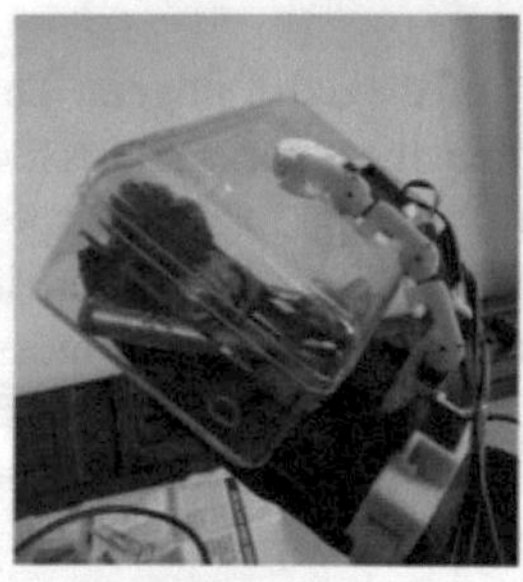

(d) Grasping box: fully actuated finger in second thumb configuration.

(e) Opening bottle with one hand: underactuated figer.

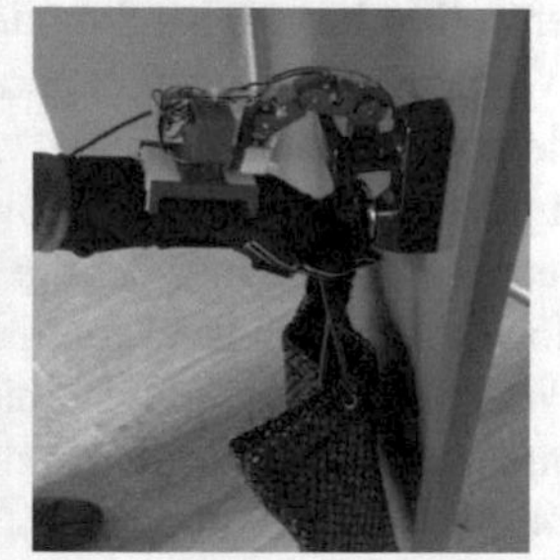

(f) Opening the door, while keeping a bag: fully actuated finger in central configuration.

**Fig. 2.9** Different tasks performed with the aid of robotic-extra fingers

In Fig. 2.9a, b, the fully and the under actuated fingers are performing ulnar grasps. In both cases the thumb, index and medium fingers grasp one object, while ring, pinkie and sixth finger, the other. In Fig. 2.9c, d the extra finger is used to perform anatomically impossible grasps. In particular, in Fig. 2.9c the second thumb configuration is used, while in Fig. 2.9d the central. The last two pictures in Fig. 2.9 shows two possible applications of the sixth finger. In Fig. 2.9e the user can unscrew a cap from a bottle using only one hand. Ulnar grasp is used to keep firm the bottle, while the other fingers can unscrew the cap. Finally, in Fig. 2.9f the fully actuated finger in the central configuration is used to open a door. The user is holding an heavy bag so cannot open the door with the same hand. He/she uses the extra finger to turn the handle, while keeping the bag with the hand.

In all the experiments, the joint reference angles of the extra-finger are related to the human hand posture. To track the human hand, we used the Cyberglove III System [67].

## 2.6 Conclusion

In this chapter, we presented a mapping algorithm to control robotic devices that can be used as extra-fingers. The control signals are computed without requiring explicit commands by the human user, but interpreting human hand motion. The number of tracked fingers in the human hand can be selected according to the task to be fulfilled so to leave some fingers free to move without interfering with the devices' behavior. The chapter also presented the design guidelines and prototype development of robotic-extra fingers for human hand enhancement. A fully and an under actuated fingers are designed and their prototypes are realized by using 3D printer and open source arduino platform. The fingers are customized for different positions on the wrist and can be worn by a rubber band. This solution allows to enlarge the hand workspace, increasing the grasp capability of the user. The control signals are computed without requiring explicit commands by the human user, but interpreting human hand motion.

We provided several experiments to prove the usability of the extra fingers. The underactuated finger resulted to be more light and portable than the other. The passive compliance of the finger allowed to adapt more easily to the shape of the grasped object. On the other side, the fully actuated version resulted to be more precise in following the high level control inputs and more powerful in terms of grasp tightness.

We believe that these simple and cost effective devices could have a great impact in understanding the effectiveness of robotic extra-fingers in augmenting human hand through wearable robotics.

# Chapter 3
# Compensating Hand Function in Chronic Stroke Patients Through the Supernumerary Robotic Finger

*The human condition today is better than it's ever been, and technology is one of the reasons for that*

*Thomas Leo Clancy*

A novel solution to compensate hand grasping abilities is proposed for chronic stroke patients. The goal is to provide the patients with a wearable supernumerary robotic finger that can be worn on the paretic forearm by means of an elastic band. The proposed prototype is a modular articulated device that can adapt its structure to the grasped object shape. The extra-finger and the paretic hand act like the two parts of a gripper cooperatively holding an object. We evaluated the feasibility of the approach with four chronic stroke patients performing a qualitative test, the Frenchay Arm Test. In this proof of concept study, the use of the supernumerary robotic finger has increased the total score of the patients of 2 points in a 5 points scale. The subjects were able to perform the two grasping tasks included in the test that were not possible without the robotic extra-finger. Adding a robotic opposing finger is a very promising approach that can significantly improve the functional compensation of the chronic stroke patient during everyday life activities.

## 3.1 Introduction

Stroke is a leading cause of long-term disabilities, which are often associated with persistent impairment of an upper limb [27]. Findings of available prospective cohort studies indicate that only 5–20% of stroke patients with a paretic upper limb manage to fully recover six months after the stroke [32]. However, 33–66% show no





I. Hussain and D. Prattichizzo, *Augmenting Human Manipulation Abilities with Supernumerary Robotic Limbs*, Biosystems & Biorobotics 26,
https://doi.org/10.1007/978-3-030-52002-1_3

recovery of upper limb functions after the same period [68, 69]. A key role in functional recovery of stroke patients with a paretic upper limb seems to be played by the improvements of the paretic hand [70, 71]. The recovery of hand functionality also contributes to a better independence in the Activity of Daily Living (ADL). In this respect, the rehabilitation projects, by which the rehabilitation team customized to the patient typologies and intensity of interventions, have been recently integrated with robotic-aided therapies. Such treatments represent a novel and promising approach in rehabilitation of the post-stroke paretic upper limb. Several robotic devices have been developed to provide safe and intensive rehabilitation to patients with mild to severe motor impairments after neurologic injury [72, 73]. The use of robotic devices in rehabilitation can provide high-intensity, repetitive, task-specific and interactive treatment of the impaired upper limb and can serve as an objective and reliable means of monitoring patient progress [74–76]. Focusing on the hand therapy, Lum et al. reviewed in [31] several works on robot-assisted approaches to motor neurorehabilitation highlighting the prototypes used in clinical tests. In [77], the authors presented a comprehensive review of hand exoskeleton technologies for rehabilitation and assistive engineering, from basic hand biomechanics to actuator technologies. However, the majority of these devices are poorly wearable, and designed to increase the functional recovery in the first months after stroke, when, in some cases, biological restoring and plastic reorganization of the central nervous system take place. To the best of our knowledge, few devices have been designed to actively compensate hand grasping function when patients are in a chronic state. In fact, when in the paretic upper limb the motor deficit is stabilized, the rehabilitation consists mainly in ergotherapy, which aims primarily in teaching compensatory strategies by using the non-paretic upper limb and by using commercial special aids [78]. This potentially increases the functional disparity between the impaired and the unaffected upper limb [33, 79].

This work focuses on the compensation of hand function in chronic stroke patients. Our idea is to augment the functional abilities of the patient with an additional robotic finger prosthesis, the supernumerary robotic finger, that is worn on the wrist/forearm of the patient, as in Fig. 3.1. Such robotic finger is used together with the paretic hand/arm to constrain the motion of the object to be grasped. The system acts like a two-finger gripper, where one finger is represented by the supernumerary robotic finger while the other by the patient paretic limb. The final aim of the supernumerary robotic finger is to increase upper limb functionality and subsequent independence in ADL in stroke survivors in relation with their personal demands and lifestyle, when no more functional motor improvements seem to be achievable. The area of impact may be identified by the "International Classification of Functioning, Disability and Health (ICF)" [80], domain "fine hand use (d 440)", "performance" in "Activity and Participation" component.

The design of the proposed compensatory tool has been driven by robotic and rehabilitation teams, starting from patients requirements in improving upper limb functionality, when the motor deficit is unchangeable. This need is particularly felt by young and social-active patients, for achieving better independence, and for continuing rehabilitation in a compensatory perspective. A preliminary version of the

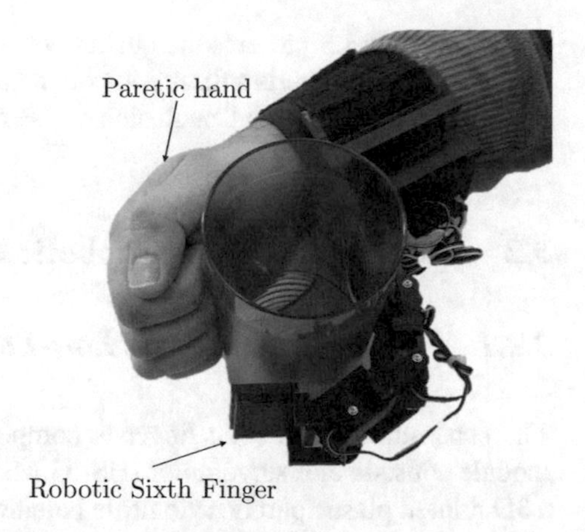


**Fig. 3.1** The supernumerary robotic finger concept. An additional robotic finger is worn by the patient on the forearm and is used in grasping tasks

device together with preliminary usability tests with healthy subjects have been presented in [25]. In [23], we have introduced an object-based mapping algorithm to control the motion of the extra robotic finger. The main feature of the mapping was to create a synergy between the human hand posture and the robotic extra-limb configuration. Such approach required to track the human hand motion to generate suitable trajectories for the device. Thus, it could not be exploited in post-stroke patients due to the very limited residual mobility of the hand. One of the challenging questions for the supernumerary robotic finger as tool for stroke patients is how to control its motion.

In this work, we propose a novel approach for the user interface which is explicitly designed for post-stroke patients. The user can control the finger flexion/extension through an electromyographic (EMG) signal captured by surface electrodes placed on the user's frontalis muscle. We have also redesigned the device to be used with this controller. With respect to the device presented in [25], we have added two new features: autonomous adaptation to the object shape and detection of patient request to tight the grasp during a manipulation task. Such dexterity is obtained through the active control of the stiffness at the joints level. The extra-finger is compliant while making the grasp with the object and becomes stiffer when more grasp force is required by the user. The admittance control implemented will be presented in Sect. 3.2.

To test the usability of the proposed prosthesis for grasp compensation, we set up a pilot experiment involving four subjects in a chronic state. The subjects worn the supernumerary robotic finger and were asked to perform the tasks comprised in the Frenchay Arm Test [81]. This test included different manipulation tasks that patients were not able to perform, but were successfully accomplished with the aid of the robotic extra-finger.

The rest of the paper is organized as follows. In Sect. 3.2 the supernumerary robotic finger prototype together with the adopted control strategies are described in details. Section 3.3 presents the qualitative test performed and the achieved results. Finally, Sect. 3.4 reports a discussion on the usability and limitation of the proposed device, while in Sect. 3.5 Conclusion and Perspectives.

## 3.2 The Supernumerary Robotic Finger

### *3.2.1 Device Structure and Low-Level Control*

The supernumerary robotic finger is composed by four one-DoF modules. Each module consists of a servomotor (HS-53 Microservo, HiTech, Republic of Korea), a 3D printed plastic part (acrylonitrile butadiene styrene, ABSPlus, Stratasys, USA) and a soft rubber part used to increase the friction at the contact area. Modules are equipped with a Force Sensing Resistor (FSR) (408, Interlink Electronics Inc., USA) placed under the rubber part and able to measure the normal component of the force applied on the module surface. The modules are connected so that one extremity of each module is rigidly coupled with the shaft of the motor through screws, while the other has a pin joint acting as revolute joint. The CAD exploded view of the modular finger and the real prototype are shown in Fig. 3.2. Technical details of the device are reported in Table 3.1.

The module connection results in a pitch-pitch configuration, which replicates the flexion/extension motion of the human finger. In [25] also abduction/adduction motion was obtained by means of a dedicated module placed on the finger base. In

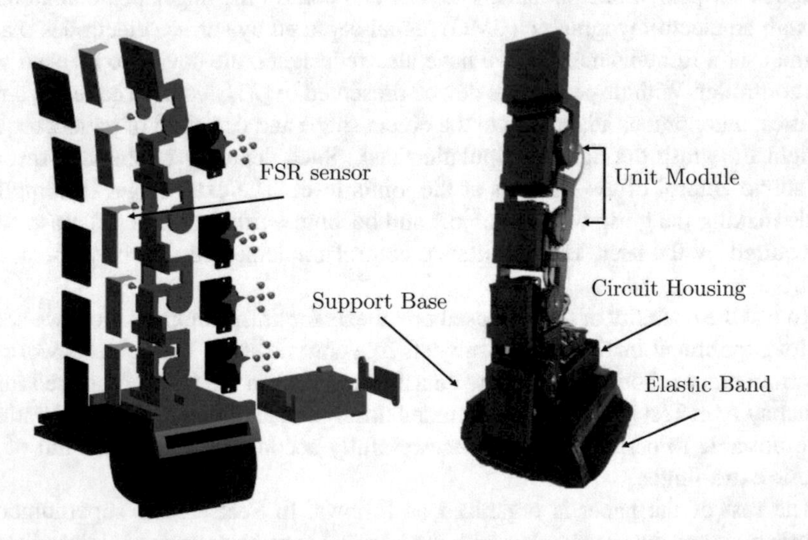


**Fig. 3.2** The supernumerary robotic finger. On the left the CAD exploded view, while on the right the prototype used in the experiments

**Table 3.1** The supernumerary robotic finger technical details

| | |
|---|---|
| Module dimension | 42 × 33 × 20 mm |
| EMG board dimension | 67 × 55 × 31 mm |
| Support base dimension | 78 × 24 × 5 mm |
| Module weight | 8 g |
| EMG board weight | 98 g |
| Support base weight | 18 g |
| Max torque per motor | 0.15 Nm |
| Stall current | 440 mA |
| Velocity of one module | 0.2 rad/s |
| EMG Board power supplies | 5 V, 3.3 V |
| Device external batteries | 7.5 V, 2.2 Ah |
| Continuous operating time | 3.5 h (@stall torque) |
| Max device payload | 610 g |

this work this possibility will not be considered. The modular part of the finger is connected to a support base which contains also the electronic housing. An elastic band allows to easily wear the device on the forearm.

An external battery is used to provide power to all the circuits. All the electronics is enclosed in a 3D printed box attached to the support base to preserve the extra-finger wearability. The module actuators are PWM controlled servomotors. The PWM signals are generated by an Arduino Nano board [82].

The motion of the supernumerary robotic finger is controlled by the user through the EMG interface, as it will be described in Sect. 3.2.2. Two possible flexion trajectories can be selected in order to grasp objects with different sizes and shapes. The two possible arising grasps have been defined as precision and power grasp, see Fig. 3.3. In precision grasps, the target is to hold the object between the paretic limb and the device fingertip pad. To this aim, the fingertip is kept parallel to the paretic limb during flexion motion. In power grasps, each module flexes with a fixed step size in order to wrap the finger around the object. The patient selects with the EMG interface the type of grasp according to the object size or the task to be executed. When the grasping phase is started, the extra-finger flexes according to the type of grasp selected. We consider the completely extended finger as the starting position to enlarge the set of possible graspable objects. The finger closing velocity is a priori set by using Arduino servo library. We set the velocity to 0.2 rad/s to let the patient reorient the device during flexion, if needed. The force sensors on the modules are used to detect contacts with the grasped object. We simplify the non-linear relation between the voltage variation on the sensors and the equivalent applied force considering a piecewise-linear function. Vernier dynamometer (Vernier, USA) was used to calibrate and verify the output of the sensors. A module is in contact with the object when the force measured with the FSR reaches a predefined threshold.

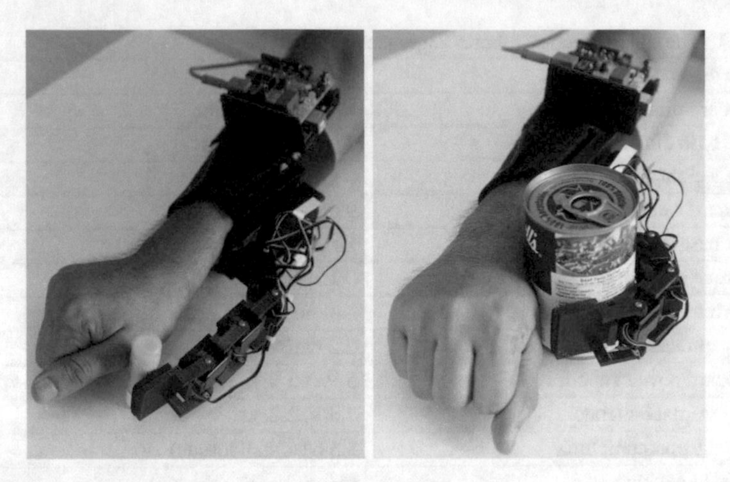

**Fig. 3.3** The two grasp modalities of the robotic extra-finger. On the left, a precision grasp is obtained between the fingertip of the device and the radial aspect of the thenar eminence. On the right, a power grasp is obtained between the whole device and the user wrist

In precision grasps, the contact is expected to occur between the object and the fingertip module. In power grasps, in order to obtain suitable contact points, we set different closing priorities according to the position of the module in the finger. If the fingertip module comes in contact first, the remaining modules stop. If another module comes in contact first, modules below to it stop, while the module above keeps moving.

When the grasping phase is complete, it is necessary to control the force exerted by the device on the object to guarantee the stability of the grasp. The forces contributing to the grasp stability are the result of the action of the paretic limb together with the action of the supernumerary robotic finger. We designed a controller that is able to regulate the force exerted by the extra-finger according to that required by the user. When the patient wishes to tight the grasp, he or she pushes the object toward the device with his or her hand/wrist. This action results in a force variation measured by the FSR placed in each module and in a displacement of the servomotors from the position reached at the end of the grasping phase.

We introduce a parameter $k_d$ to regulate the position error of the servomotor proportional to force observed by FSR. We set a linear relation between the force variation measured by the FSR placed in each module and the value of $k_d$. In particular, the range of forces read by the force sensors (0.6–6 N) was linearly mapped in the range 0.3–3 of parameter $k_d$. Then, the possible angle displacement $\Delta q$ is computed as

$$\Delta q = k_d(q_{des} - q_m),$$

where $q_m$ is the actual position of the servomotor while $q_{des}$ is its desired position. We modified the servomotor in order to measure its actual position by accessing the internal encoder measurements. The only servomotor parameter that can be set is its desired position $q_{des}$. At time instant $t$ the desired position for the $i$th servomotor is computed as

$$q_{des,i}(t) = q_{m,i}(t-1) + \Delta q_i(t-1).$$

In presence of a rigid grasped object, the measured positions of the extra-finger joints do not change due to the object constraints. So that, changing the desired position of the servomotors we can control the force exerted by the device onto the object.

The resulting behavior is similar to an admittance control of the motor. Generally, in admittance control framework, the compliant (or stiff) behavior of the joints is achieved by virtue of the control, differently from what happen in mechanical systems with a prevalent dynamics of elastic type [83]. Varying the parameter $k_d$ the module appear to be more or less stiff with respect to an external applied force.

We set priorities between modules for the simulated compliance variation, similarly to what we did for the grasping procedure. So that, if, for instance, only the fingertip module is in contact with the object, all the other modules change their stiffness accordingly. This solution allows to control the stiffness of modules that are not in contact with the object.

When the patient wants to release the grasped object, he or she needs to lower the force exerted by his or her hand/wrist on the object so to decrease the value of $k_d$. Eventually, using the EMG interface, the robotic finger can be placed back to its home position by following a predefined trajectory.

Note that if the patient is not able to exert force on the object, due to his or her motor deficit or to the position of the supernumerary robotic finger on the forearm, the grasp tightness can be controlled through the EMG interface. The patient can activate the finger flexion, wait for finger shaping and then activate again the flexion to tight the grasp. This solution requires a higher cognitive effort from the patient that has to estimate which is the sufficient grasp force necessary to hold the object. When possible, admittance control is preferable also in order to optimize battery consumption.

### *3.2.2 EMG User Interface*

The interface with the user must be designed so to appear intuitive and simple. As a preliminary solution, the patient could activate the robotic extra-finger flexion/extension through a switch represented by a push button embedded on a ring worn on the healthy hand [26, 36]. Although this solution dramatically simplified the interaction with the device, wearing a ring on the contralateral hand and operating the switch could prevent the healthy hand from doing other tasks. For example, when performing bimanual tasks or when grasping an object, the subject had to be careful not to activate the switch unintentionally. For this reason, we have developed an EMG interface which maintains the principle of simplicity of the switch, but that leaves the patient free to use his or her healthy hand.

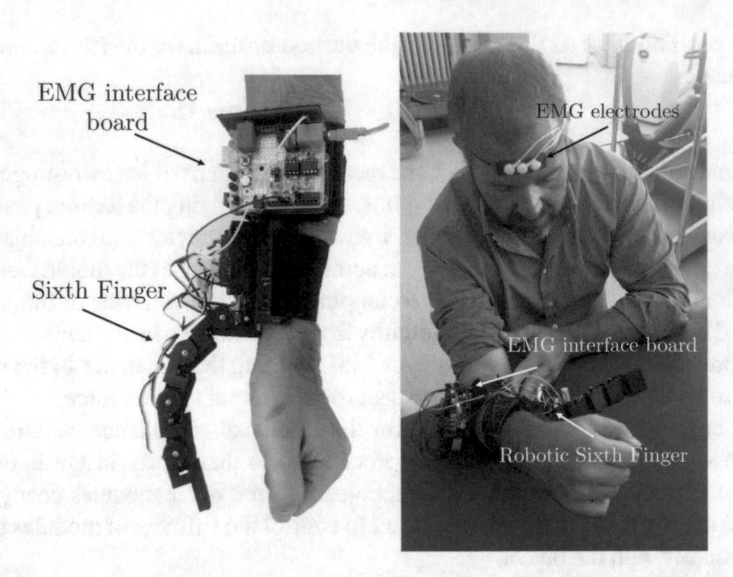


**Fig. 3.4** On the left, the EMG board used. On the right the EMG electrodes placed on the patient head

Thanks to the proposed EMG interface, patients can consciously control the supernumerary robotic finger by contracting the frontalis muscle on their forehead. Bipolar EMG electrodes are positioned on the frontalis muscle, since its functionality, due to a bilateral cortical representation, is always spared in case of a hemispherical stroke. Activation of the muscle is achieved by moving the eyebrows upward. Functionality of muscles of the paretic limb is usually sub-optimal, hence scarcely reliable for EMG control. Moreover, using the paretic limb for a dual task (i.e., EMG control and grasping) can be too demanding for most of patients. In the proposed system, patients worn a headband where surface electrodes are placed, see the right side of Fig. 3.4. The electrical activity measured by the electrodes (one ground and two recording electrodes for true differential recordings) is acquired through an embedded system (Muscle SpikerShield, Backyard Brains, Ann Arbor, MI - USA) that can be worn on the patient's arm thanks to an elastic band, see the left side of Fig. 3.4. This EMG interface system for bio-signals acquisition is designed to be used with dry electrodes and no skin preparation. The board includes a gain knob to tune the amplification of the muscle recording depending on the size and type of muscle. This feature has been used to filter involuntary muscle contractions. The EMG interface board consists of a signal conditioning board and an Arduino Uno that processes the signals. Dimensions and weight of the board are reported in Table 3.1. We have designed the processing unit in order to obtain a trigger signal if the recorded EMG signals exceed a pre-set threshold.

The EMG interface is thus able to detect if the frontalis muscle of the patient is voluntary contracted. Upon this, we have implemented a high level control strategy that is summarized in Table 3.2. With one muscle contraction the patient controls

**Table 3.2** Commands detected through the EMG interface

| EMG signal | Associated action |
|---|---|
| One contraction | Move/stop |
| Two contractions | Change direction |
| Three contractions | Change grasp mode |

the motion/stop of the finger. When the finger is stopped, two contractions in a time window of 1.5 s switch the motion direction from flexion to extension and viceversa. Finally, when the finger is in its starting position (completely open), the user can change the finger flexion trajectory (power or precision grasp) making three contraction in a time window of 2 s.

### 3.2.3 Device Positioning in the Patient Paretic Limb

The position of Robotic Sixth Finger on the patient forearm is an important aspect to take into account. Before deciding where to wear the device, it is necessary to study both the residual mobility of the patient and the ADL we want to accomplish. The Robotic Sixth Finger can be placed in the distal part of the forearm (near, or on the wrist) since the grasp is obtained by opposing the device to the paretic hand.

However, the distal positioning of the supernumerary robotic finger may fail when the post-stroke motor deficit is so advanced that a pathological synergism in flexion has taken place. In this case, the wrist becomes too much flexed and fingers are too much closed towards the palm thus not allowing a successful grasping. When this pathological condition occurs, the supernumerary robotic finger may be positioned more proximal at the forearm, in a way that the grasp can be achieved by the extra-finger opposition to the radial aspect of the thenar eminence or to the anatomical snuff box. An example of two possible positions for the supernumerary robotic finger are reported in Fig. 3.5.

This flexibility in the positioning is achieved thanks to the modularity of the structure and the flexibility of fixing support. Modularity allows to regulate the size and dexterity of the finger according to the position on the forearm and according to the limb characteristic of each patient. The regulation is obtained adding/removing modules on/from the device. The support base of the finger can be translated or rotated along the arm to place the finger on a suitable position and orientation. An elastic band and rubber spacers are used to increase the grip and comfort, while reducing the fatigue during continuous use of the finger.

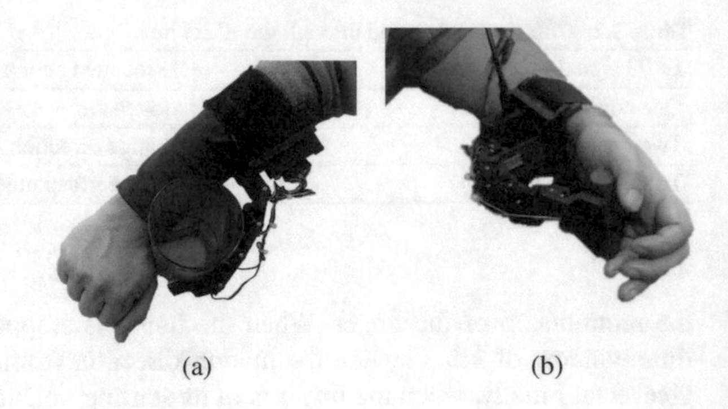


**Fig. 3.5** The robotic extra finger in two possible configurations on the forearm. **a** The grasp is obtained at the wrist level, **b** The grasp is obtained at the hand level

## 3.3 Pilot Experiment

In the current proof of concept study, we tested with four subjects (three male, one female, age 48–60) how the supernumerary robotic finger can compensate for grasping capability. The aim was to verify the potential of the approach and to understand how rapidly the subjects can successfully interact with the wearable device. In this direction, we performed a fully ecological qualitative test, the Frenchay Arm Test [81] as shown in Fig. 3.6. The test consisted of five pass/fail tasks to be executed in less then three minutes. The patient scored 1 for each of the successfully completed task, while he or she scored 0 in case of fail. The subject sat at a table with his hands in his lap, and each task started from this position. He or she is then asked to use the affected arm/hand to:

1. *Task_1* Stabilize a ruler, while drawing a line with a pencil held in the other hand. To pass, the ruler must be held firmly (see Fig. 3.6a).
2. *Task_2* Grasp a cylinder (12 mm diameter, 50 mm long), set on its side approximately 150 mm from the table edge, lift it about 300 mm and replace without dropping (see Fig. 3.6b).
3. *Task_3* Pick up a glass, half full of water positioned about 150 to 300 mm from the edge of the table, drink some water and replace without spilling[1] (see Fig. 3.6c).
4. *Task_4* Remove and replace a sprung clothes peg from a 10 mm diameter dowel, 150 mm long set in a 100 mm base, 150–300 mm from table edge. Not to drop peg or knock dowel over (see Fig. 3.6d).
5. *Task_5* Comb hair (or imitate); must comb across top, down the back and down each side of head (see Fig. 3.6e).

When compared with other upper limb assessments, the Frenchay arm test has shown good reliability in measuring functional changes in stroke patients [81]. The score ranges from 0 (no one item performed) to 5 (all the items performed). All the subjects taking part to the experiment have been affected by stroke more than two years before the test. The rehabilitation team has declared that no more functional improvements

[1]Note that for safety reasons we did not use water in presence of electronic components.

(a) Task 1: stabilize a ruler, while drawing a line with a pencil held in the other hand. Note that the supernumerary robotic finger does not interfere with the task execution.

(b) Task 2: grasp a cylinder (12 mm diameter, 50 mm long). A precision grasp is used to grasp the small cylinder.

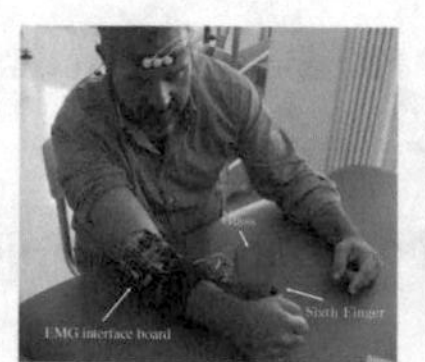


(c) Task 3: pick up a glass and drink. A power grasp strategy is used to achieve the grasp.

(d) Task 4: remove and replace a sprung clothes peg from a dowel.

(e) Task 4: remove and replace a sprung clothes peg from a dowel.

**Fig. 3.6** The frenchay arm test

are achievable with respect to the gained upper limb motor performance. The supernumerary robotic finger can be used by subjects showing a residual mobility of the arm. For being included in the pilot experiment, patients had to score ≤2 when their motor function was tested with the National Institute of Health Stroke Scale (NIHSS) [84], item 5 "paretic arm". Moreover, the patients showed the following characteristics: (1) normal consciousness (NIHSS, item1a, 1b, 1c = 0), absence of conjugate eyes deviation (NIHSS, item 2 = 0), absence of complete hemianopia (NIHSS, item 3 ≤ 1), absence of ataxia (NIHSS, item 7 = 0), absence of completely sensory loss (NIHSS, item 8 ≤ 1), absence of aphasia (NIHSS, item 9 = 0), absence of profound extinction and inattention (NIHSS, item 11 ≤ 1). Patients received the supernumerary robotic finger in the paretic hand, the left hand for two subjects and the right one for the other two. Thanks to the device design, the same prototype can be worn either on the right or on the left arm. Written informed consent was obtained from all participants. The procedures were in accordance with the Declaration of Helsinki.

The rehabilitation team assisted the subjects during a training phase that lasted about one hour. During this phase, the optimal position of the device on the arm according to the patient motor deficit was evaluated. The patients also tried the EMG interface in order to become confident with the extra-finger high-level control. Two patients tried the extra-finger for the first time. After the training, the subjects had three minutes to perform the Frenchay Arm Test. All the subjects performed the test twice, one with and one without the device. We randomly decided the starting condition. The supernumerary robotic finger was placed on the paretic limb, while it was activated using the EMG interface. In Fig. 3.6a–e snapshots of the execution of

**Table 3.3** Results of the Frenchay Arm Test (FAT) for the four patients with and without using the supernumerary robotic finger

| Task | Patient 1 | | Patient 2 | | Patient 3 | | Patient 4 | |
|---|---|---|---|---|---|---|---|---|
| | With | Without | With | Without | With | Without | With | Without |
| Stabilize a ruler | 1 | 1 | 1 | 1 | 1 | 1 | 1 | 1 |
| Grasp a cylinder | 1 | 0 | 1 | 0 | 1 | 0 | 1 | 0 |
| Pick up a glass | 1 | 0 | 1 | 0 | 1 | 0 | 1 | 0 |
| Remove a sprung | 0 | 0 | 0 | 0 | 0 | 0 | 0 | 0 |
| Comb hair | 0 | 0 | 0 | 0 | 0 | 0 | 0 | 0 |

**Table 3.4** Questionnaire and relative marks. The mark ranges from "0 = totally disagree" to "7 = totally agree". Mean and standard deviation (Mean (STD)) are reported

| Question | Answer |
|---|---|
| The EMG interface results intuitive easy to use | 6.75(0.5) |
| I did not need any particular training to start using the interface | 6(0.81) |
| I felt confident using the system | 6.25(0.5) |
| I think that I would need the support of a technical person to be able to use this system every day | 3.25(1.71) |
| The system was easy to use | 6.25(0.5) |
| I would imagine that most people would quickly learn how to use this system | 7(0) |

the five tasks are reported, respectively. The results of the test are shown in Table 3.3 for the four patients. All the subjects selected a precision grasp for the cylinder, while a power grasp was used for the glass.

Task 2 and Task 3 were successfully accomplished by all the patients with the help of the extra-finger. Task 4 and Task 5 involved the grasp of an object and a rather complex manipulation part. Although the patient were able both to grasp the sprung and the comb, the poor residual mobility of the arm caused a failure in the task fulfillment. Task 1 was executed both with and without the device. This means that the device did not interfere with the residual mobility of the paretic limb thank to its high wearability and portability.

After the Frenchay Arm Test, the patients were asked to fill a questionnaire in order to evaluate the subjective feelings on the ease of use of the device. Results are reported in Table 3.4. Moreover, the two patients that already tried a previous version of the device with a switch-ring interface declared to prefer the new EMG interface.

## 3.4 Discussion

The exploitation of the supernumerary robotic finger in post-stroke compensation of hand functions is at an early stage. The device has been built upon the ideas of wearability, ergonomics, safety, ease of use and comfort, while increasing the upper limb functionality of post stroke patients. Most of the devices designed for hand rehabilitation share the aim of assisting finger motion during an assigned training. Many algorithms for human-robot interaction can be programmed also into simple robotic devices. Lum et al. have classified in [31] the main features of different control strategies available in literature. Although our device is not designed for hand rehabilitation but for compensating hand function in grasping tasks, some of the proposed strategies has been taken as guidelines for the design and interface of our devices. In particular, we considered: (i) adaptation to patient performance; (ii) prevention of fatigue and frustration and (iii) production of more physiologically accurate movement patterns. The adaptation to the patient performance is obtained considering the versatility of the device to be worn in different parts along the forearm according to patient capability of oppose to the device motion. Moreover, the control of the compliance of the device allows the user to select the tightness of the grasp. To prevent fatigue and frustration on the patient, the motion of the supernumerary robotic finger is not mechanically coupled with the human limbs. The patients' muscle are not constrained to follow any particular trajectory imposed by the extra-finger. Patients only need to be able to place their forearm in a suitable position, so to let the object be in the workspace of the robotic device.

Finally, in case of traditional exoskeletons, precautions need to be carried out in terms of kinematics design and during their use to avoid any abnormal posture that can produce any physiologically inaccurate movement patterns. The use of the extra-finger results to be safer in this regard.

All the patients were able to improve their performance in tasks of everyday activities without the need of any specific training period, thus suggesting that the successful use of the supernumerary robotic finger is rather intuitive, at least for these very basic activities. This aspect is also confirmed by the answer to the questionnaire in Table 3.4. The ease of use represents an important point, taking into consideration that a proportion of stroke patients may also complain of some cognitive deficits, possibly limiting their compliance during a demanding learning phase. Now, the challenge is to identify and better understand the different requirements coming from the different application scenarios.

The pathological synergism in flexion is characterized by the arm and the forearm intrarotated, adducted and flexed at their main joints, as well as fingers closed. Patients may show a deficit not only in grasping the object and keeping it, but also in releasing it. This motor pattern is mainly due to two factors: weakness of extensors muscles and hypertonia of flexors. The patients that have tried the supernumerary robotic finger experienced a new ability in releasing the utilized objects without the aid of the healthy hand, as well as the ability to grasp an object without the help of the contralateral limb.

Compensation process by using extra-finger motivates the patient to use her or his muscles to coordinate with the device for the completion of the task. Thus, the extra-finger acts like an active and motivational assistance device. This approach encourages the patients to use their potential and residual abilities effectively instead of being fully dependent on the motion of robotic device like passive assistive devices [85]. In this view, the use of the supernumerary robotic finger shares conceptual similarities with the constraint-induced movement therapy (CIMT), a rehabilitative approach characterized by the restraint of the healthy upper limb accompanied by the shaping and repetitive task-oriented training of more affected upper extremity, with the purposes of overcoming the learned nonuse phenomenon of the hemiplegic upper extremity [86]. However, the obvious advantage of the supernumerary robotic finger is that there is no need to immobilize or restraint the healthy limb to favor the (re)use of the paretic hand.

## 3.5 Conclusion and Perspectives

In this chapter we presented a robotic compensatory tool that can be used by chronic stroke patients to regain grasping capabilities at the paretic hand. The latter phase of post-stroke rehabilitation is identified by the "compensation phase". In this phase, functional recovery is based on the learning of newly acquired motor strategies to compensate the neurological deficit. These strategies may sometimes be neither ergonomic nor ecological, or may even increase pathological motor patterns, usually by worsening tonic flexion at the forearm of the paretic limb. We expect to increase patients' performances, with a focus on objects manipulation, thereby improving their independence in ADL, and simultaneously decreasing erroneous compensatory motor strategies for solving everyday tasks [87].

The proposed control of the robotic finger dramatically simplifies the human robot interaction since only the activation of a grasping procedure through the EMG interface is needed.

# Chapter 4
# Design and Development of Soft Supernumerary Robotic Fingers for Grasp Compensation in Chronic Stroke Patients

*I can never satisfy myself until i can make a mechanical model of a thing. If I can make mechanical model, I can understand it*

*Lord Kelvin*

This chapter presents design, analysis, fabrication, experimental characterization and evaluation of two prototypes of soft supernumerary robotic fingers that can be used as grasp compensatory devices for hemiparetic upper limb.

In recent years, several researchers have focused on the development of simple, compliant, yet strong, robust, and easy-to-program manipulation systems to overcome the common issues of rigid multi-fingered robotic hands. Although many prototypes have been proposed, there is still a lack of a systematic way for soft hand design. One of the most critical element that plays the role in the successful grasp and shape adaptation of the object is the trajectory of the robotic fingers which can be regulated by acting on their joints stiffness. We propose a method to compute the stiffness of flexible joints and its realization in order to let the fingers track a certain predefined trajectory. We refer to tendon driven, underactuated and passively compliant hands composed of deformable joints and rigid links. The finger joints can be given specific stiffness and pre-form shapes such that a single cable actuation can be used. We define firstly a procedure to determine suitable joints stiffness and then we propose a possible realization in robotics fingers hardware structure. The stiffness computation is obtained leveraging on the the mechanics of tendon-driven hands and on compliant systems, while for its implementation beam theory has been exploited. We validate the proposed framework both in simulation and with experiments using a prototypes of the devices.



I. Hussain and D. Prattichizzo, *Augmenting Human Manipulation Abilities with Supernumerary Robotic Limbs*, Biosystems & Biorobotics 26,
https://doi.org/10.1007/978-3-030-52002-1_4

The devices are the results of experimental sessions with chronic stroke patients and consultations with clinical experts. Both devices share a common principle of work which consists in opposing to the paretic hand/wrist so to restrain the motion of an object. They can be used by chronic stroke patients to compensate for grasping in several Activities of Daily Living (ADL) with a particular focus on bimanual tasks. The robotic extra fingers are designed to be extremely portable, wearable, robust and capable to adapt to different object shapes. They can be wrapped as bracelets when not being used, to further reduce the encumbrance. Both devices are intrinsically-compliant and driven by a single actuator through a tendon system.

## 4.1 Introduction

Wearability is the key feature of the robotic devices designed for assisting the patients in activity daily living, since the devices should be used by the patients also outside the rehab facilities. In [9], Pons et al. outlined two possible categories of wearable robots: exoskeletons and prosthetic robots. The former are designed to complement the ability of the human limb and restore the handicapped function usually mapping onto the anatomy of the human limb. The latter are electro-mechanical devices that substitutes for lost limbs after amputation. In this work, we enrich these definitions by introducing a wearable robot which is grounded on the human body, but that is not mechanically coupled with the human limb. We do not attempt to assist the paretic hand/arm motion of the patient, but rather we add just what is needed to grasp: an extra thumb. The robotic extra-finger is worn on the user forearm and can accomplish a given task in cooperation with the paretic limb. A preliminary prototype, called the Robotic Sixth Finger, has been proposed in [25, 26, 36]. The fully actuated structure allows the Robotic Sixth Finger to actively shape around an object, but the resulting size and weight affect the device portability and wearability.

In this work, we present two novel prototypes of wearable grasp compensatory devices for hemiparetic upper limb: the soft sixth finger and the double soft sixth finger, see Fig. 4.1. The devices have been designed to guarantee high wearability and portability. The devices can be worn as a bracelet when these are not being used. The patient can switch from the bracelet to the operative position by using his their healthy hand. Moreover, in the previous versions of the device, the batteries and all the electronic boards for the control were worn in the paretic arm by means of an elastic band. The arising weight limited the mobility on the patient's paretic arm. In the Soft-SixthFinger, we moved all the batteries and the electronics board from the paretic arm to a box that can be fixed in the user belt. The result is a sensible reduction of the weight the user has to support with the paretic arm. The double soft sixth finger device has been designed to improve grasp stability in more payload demanding tasks. We doubled the flexible structure of the device to obtain two fingers. The two fingers are attached to a base that can be worn at the user's forearm. The device is actuated by a single motor and the two fingers design improves the grasp stability and robustness. we presented a systematic way to compute the stiffness

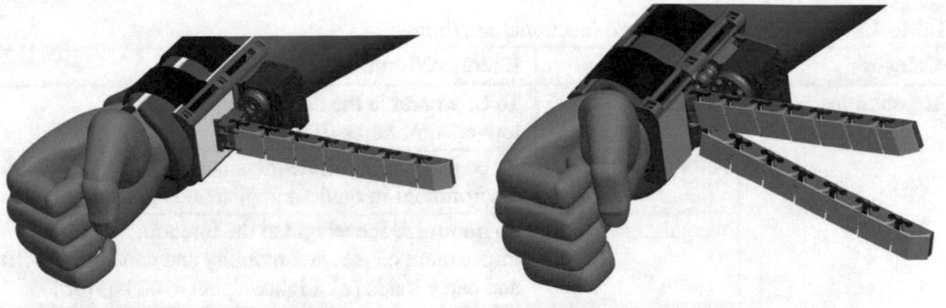

(a) The soft sixth finger concept. (b) The double soft sixth finger concept.

**Fig. 4.1** The proposed wearable robotic compensatory devices for hemiparetic upper limb. **a** The soft sixth finger: a single extra finger is proposed to compensate for missing hand grasping function **b** The double soft sixth finger: a double finger is proposed to increase the grasp stability and the payload

ratio between the passive compliant joints so to obtain a desired trajectory for the finger. The proposed method assumes a given target motion of the fingertip and a given maximum actuation force for the tendon driven system to compute the stiffness value of the passive joints. We proposed a modular approach to define robotic fingers. The performances of the devices were through an experimental setup. The results showed major improvements in the performance characteristics of the devices as compared to older version of the device, in particular, the improvement in grasp stability, fingertip force and maximum payload.

## 4.2 The Soft Sixth Finger and the Double Soft Sixth Finger

The wearable assistive robotic devices used for clinical applications must meet specific human factors and performance criteria. The general guidelines which can be found in the literature include: durability, energy efficiency, low-encumbrance, ease of use, error tolerance and configurability [88]. The specific performance criteria and human ergonomics strongly depends on the actual patient conditions and needs [89, 90]. In order to improve the wearability and the acceptability of the devices for the users, the design must meet a number of conditions related to ergonomics and functionality [91]. Several experiments with the patients conducted in cooperation with a rehabilitation team reported in [2, 36, 37, 92] led us to outline the main functional requirements and human factors reported in Table 4.1.

We considered these ergonomics and functional requirements in the design and development of the proposed robotic devices. Table 4.8 summarizes how we met these requirements in our devices while the detailed design and development methodology are explained in corresponding sections.

**Table 4.1** Device Ergonomics and functional requirements

| Category | Requirement | Literature/Experimental observations |
|---|---|---|
| Ergonomics | Wearability | To be wearable the device should be of low-encumbrance [93] |
| | Portability | To be used also outside the laboratory and structured environment in outdoor applications [94] |
| | Weight | To minimize the weight at the forearm. The arm impairment causes less mobility and capability to lift and carry loads [2]. Lightweight ($< 400$ g) [35] |
| | Ease of use | To provide freedom to patients in wearing and using the device without any assistance [89] |
| Functional | Robustness | To resist to unwanted collisions with the environment [38] |
| | Fatigue avoidance and safety | To avoid un-natural forces on human muscles during the use of device which can cause fatigue and frustration [9], the human-robot safe interaction must be ensured [95] |
| | Device adaptability to patient's conditions | The device positioning according to patient's conditions and residual mobility of the arm/hand [37] |
| | Control interface adaptability to patient's conditions | The bio-signal control interface must be adaptive to patient dependent nature of signal variations and given detection conditions [96] |
| | Device coupling with human arm | To firmly couple the device with human forearm, in order to realize a stable grip even in the presence of heavy loads [97] |
| | Object shape adaptability | To adapt to objects with different shapes and sizes [98] |
| | Stable grasp | To realize a stable grasp during holding, lifting and pouring tasks, involving objects of various sizes and weights |
| | Mechanical power | To generate the contact forces (5 $N$) required for successfully grasping objects of daily living [11] |
| | Configurability | To easily adjust according to patient specific needs, e.g., the desired control interface, the length of the device, easy assembly of parts, etc. [89] |
| | Functional versatility | To accommodate a wide variety of objects and tasks associated with ADL [99] |
| | Energy efficiency | To avoid expensive, unsafe and heavy energy sources [94] |
| | Error tolerance | To provide robustness to positioning and sensing errors, and having mechanical design features suited to grasp stability including: high friction, natural distributed compliance and underactuation [98] |
| | Simple and intuitive interfaces | To realize simple and user friendly control interface [11] |

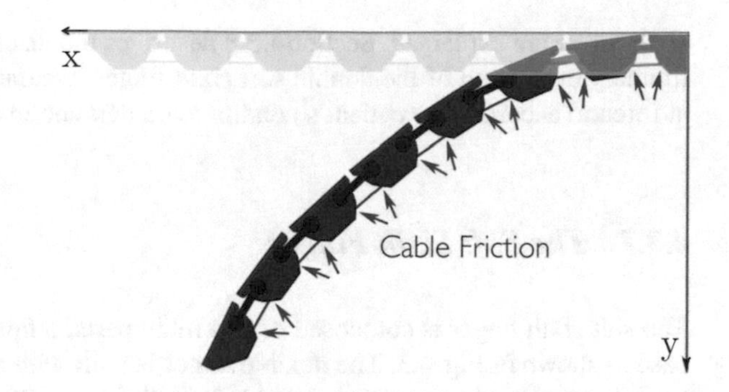


**Fig. 4.2** Underactuated cable driven flexible finger. The finger has modular structure. Each module is composed of soft and stiff parts

The compensatory devices have been designed to be wearable, robust and capable to adapt to different object shapes. In general, the robustness plays a twofold role. First, it enables the robotic devices to reliably grasp objects in the presence of large sensory uncertainty. Second, it enables the devices to withstand large impact and other forces due to the unintended contact with the environment. The robustness and soft interaction are mainly achieved by either regulating the compliance of robotic joints [100] or tuning the intrinsic softness, acting on the passive characteristics of the robot bodyware [65, 101–103]. The former approach is based on complex and bulky variable impedance actuators aiming to be used in different contest [104, 105]. Our devices are inspired by the latter approach in order to be simple, light in weight and compact in size. The proposed devices are based on principle of underactuated cable driven flexible and modular structure as shown in Fig. 4.2. The passive compliance in the joints guarantees the robustness and safety during the interaction with the environment. The devices can endure collisions with hard objects and even strikes from a hammer without breaking into pieces. In addition to this, the actuation system of the proposed devices resembles that of underactuated robotic hands [106]. Underactuated hands have desirable adaptability to shapes, and can be effectively implemented using relatively simple differential and elastic elements [107, 108]. The transmission solutions allow motion of other joints to continue after contact occurs on a coupled link, allowing the hands to passively adapt to the object shape [98, 109, 110]. Passive adaptability allows to drive the device with a reduced number of control parameters. The built-in compliant nature of the extra-fingers and underactuation increase their ability to grasp different objects [111]. They can adapt their shape to that of the grasped object. Shape adaptation increases the grasp performance by compensating the uncertainties in sensing, actuation and helps in stabilizing the grasp [112]. The robustness and intrinsic compliance is realized through the cable driven flexible structure of the robotic fingers. A mathematical model, presented in Sect. 4.3.3, has been used to study the kinematics of such cable driven flexible finger and to simulate its bending profile. The simulation results helped in minimizing the manufacturing iterations, in particular adjusting the stiffness in each joint to obtain a desired finger closing trajectory and length of finger to cover a certain workspace. In Sect. 4.2.1 the detail design, development and performance evaluation of the soft sixth finger are explained. Section 4.3.8 describes the design, development and preliminary evaluation of the double soft sixth finger. Wearability and device position at forearm according to patient's condition are detailed in Sect. 4.3.9.

### *4.2.1 The Soft Sixth Finger*

The soft sixth finger is composed of two main parts, a flexible finger and a support base as shown in Fig. 4.3. The flexible finger is built with a modular structure. Each module consist of a rigid 3D printed link realized in ABS (Acrylonitrile Butadiene Styrene, ABSPlus, Stratasys, USA) and covered with a silicon skin and a 3D printed thermoplastic polyurethane part (Lulzbot, USA) that acts as the flexible joint. We selected polyurethane for flexible parts because the high elongation of this material allows for repeated movement and impact without wear or cracking proving also an excellent vibration reduction. Reasons for adding passive elements are manifold, including storing elastic energy, avoiding tendon slackness, passive compliance, the distribution of forces over a large contact area and ensuring the uniqueness of the position [113]. The modules are connected by sliding the thermoplastic polyurethane part in the ABS part. This method makes the assembling process easy without using any screw or passive elements to combine the modules. The length and closing trajectory of the device have been selected according to the procedure presented in Sect. 4.3.

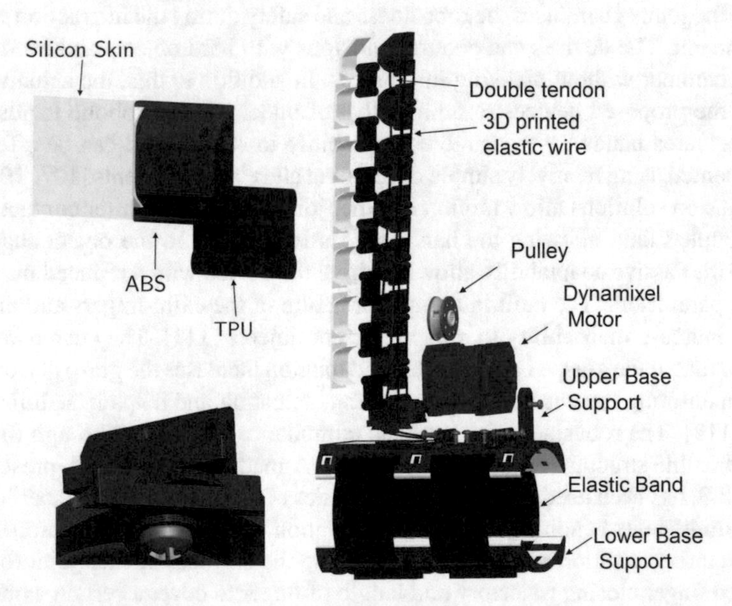


**Fig. 4.3** The CAD exploded view of unit module, complete soft sixth finger and Dovetail passive locking mechanism. Two holes in the module for double tendon, modular structure of the device, support base and actuator are shown

The device is designed and developed by combining two different manufacturing processes, i.e., 3D printing and moulding. The skeleton of the device is fabricated by rapid prototyping 3D printing while the silicon skin is realized by moulding process. The moulding process shapes the raw material using a solid frame of a particular shape, called a pattern. We used 3D printed skeletons to hold the liquid silicon in desired shape until it turned solid. We realized closed-top moulds which are used for more complex part geometries. We pored the silicon mixture over the skeletons of the modules and support base and used other mold's parts to constrain the liquid silicon to achieve the desired geometry and shape of the skin. Metal tubes were inserted into the module holes so to avoid silicon to fill the tendon holes. In Fig. 4.4, the top row shows the exploded view of the parts used in the moulding process, the bottom row shows the assembled configuration of parts during the curing process of liquid silicon. The silicon used is Fast Rubber FR-18 which is bi-component and cures at room temperature. The mixing ratio of components are 100 g of resin per 5 g of hardner. It has viscosity of 30 Pa s and the final hardness is 17±2 shore A. The silicon skin on the rigid links is realized through casting process, aiming to increase the friction at the possible contact areas. The mold parts used in casting process are shown in Fig. 4.4.

The support base of the device has been designed to assure a firm grip on the arm. The ergonomic design of the support base guarantees the stability of the device

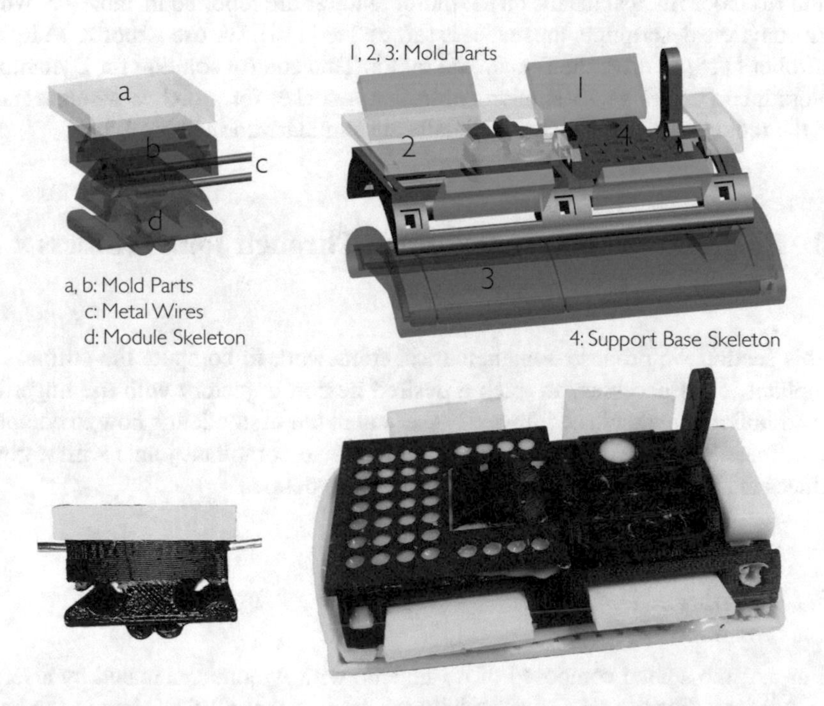


**Fig. 4.4** Moulding process to realize silicon skin. Mold parts, module skeleton and support base skeleton are 3D printed. The metal wires are inserted in the tendon wire holes

to withstand the applied load. The support base consists of two parts coupled with velcro strips to facilitate wearing the device at the forearm and guarantee adaptation to different arm sizes. The upper part contains the actuator and base module of the robotic finger. Both parts are covered with a silicon skin to increase the comfort and stability at the forearm. The skeleton of the support base is 3D printed by ABS material and silicon skin is realized through the moulding process. The support base and the flexible finger are coupled through Dovetail triangle locking mechanism, which is described in detail in Sect. 4.3.9. This mechanism is used to switch between working and rest position of the devices. The structure of the support base is symmetrical, feature that enables the robotic devices to be worn on both the left or the right arm of the patients without any modification in the device. The device actuation is achieved by using a single actuator and two tendons running in parallel through the modules of the finger. The holes in the rigid links allow the passage of the cables (polyethylene dyneema fiber, Japan) which are used to realize the tendon driven actuation. The tendon wires run through the finger and are attached on one side to the fingertip and on the other to a pulley rigidly connected to the actuator shaft. When the motor is actuated, the tendon wires are wound on the pulley reducing the length of the wire and thus flexing the finger. As the motor is rotated in opposite direction, the extension of the finger is achieved thanks to elastic force stored in the flexible parts of the modules. The actuator used is the Dynamixel MX-28T (Robotis, South Korea). Principal details on the motor features are reported in Table 4.5, while for a complete description, the reader is referred to [114]. We use ArbotiX-M Robocontroller [115] to drive the Dynamixel motor. This control solution for Dynamixel motors incorporates an AVR microcontroller, a socket for an Xbee wireless radio and the motor driver. The technical details are summarized in Table 4.5.

## 4.3 Fingertip Trajectory Definition Through Joint Stiffness Design

In this section we propose a mathematical framework to compute the stiffness of compliant joints necessary to track a desired flexion trajectory with the fingertips of a compliant underactuated finger. In particular, we firstly define how to compute the stiffness values and then how to realize passive compliant joints with a given stiffness once the geometry of the passive joints is defined.

### *4.3.1 Problem Definition*

Let us assume a hand composed of $n_f$ fingers, with $n_q$ joints, actuated by a series of $n_t$ tendons. For the sake of simplicity we assume that all the joints in the hand are revolute (R), so that the variable $q_i$ describing $i$th displacement is a rotation.

Let us indicate with $\mathbf{q} = [q_1, \cdot, q_{n_q}]^{\mathrm{T}} \in \Re^{n_q}$ a vector containing hand joint rotations and with $\mathbf{t} \in \Re^{n_t}$ tendon displacements. Let us collect in the vector $\mathbf{p} \in \Re^{n_f n_d}$ the vectors $\mathbf{p}_i \in \Re^{n_d}$ describing position (and eventually orientation) of a reference frame representing the $i$th fingertip with respect to a reference frame that we assume as a basis, fixed, for instance, on hand palm. We indicate with $n_d$ the dimension of the configuration space for the fingertip, for example, $n_d = 3$ in the planar case, $n_d = 6$ in the more general three dimensional case.

From the kinematic analysis of hand fingers, it is possible to relate tendon displacements $\mathbf{t}$ to hand joint configuration $\mathbf{q}$, i.e.,

$$\mathbf{t} = \mathbf{t}(\mathbf{q}), \tag{4.1}$$

where $\mathbf{t}(\mathbf{q}) : \Re^{n_q} \rightarrow \Re^{n_t}$ is the transformation function, that in the more general case is nonlinear and depends on system geometric dimensions and tendon routings. Considering a small variation of tendon lengths and joint angles with respect to a reference configuration, the above relationship can be linearised as

$$\delta\mathbf{t} = \mathbf{T}(\mathbf{q})\delta\mathbf{q}, \tag{4.2}$$

where $\mathbf{T}(\mathbf{q})$ is the Jacobian of the function introduced in Eq. (4.1), that can be evaluated as $\mathbf{T}(\mathbf{q}) = \frac{\partial \mathbf{t}}{\partial \mathbf{q}}$. In some cases, when the tendons are connected through pulleys with a circular radius, it can be shown that Eq. (4.1) is linear and then $\mathbf{T}$ matrix is constant. In this case $\mathbf{T} \in \Re^{n_t \times n_q}$ is a transformation matrix whose elements depend on finger pulleys' sizes and tendon routing topology and is independent from hand posture [106]. By applying the Principle of Virtual Work to the hand it is possible to obtain the dual static relationship

$$\boldsymbol{\tau} = \mathbf{T}^{\mathrm{T}}\mathbf{f}, \tag{4.3}$$

where $\boldsymbol{\tau} \in \Re^{n_q}$ is a represents hand joint torques and $\mathbf{f} \in \Re^{n_t}$ is a vector containing tendons' pulling forces. If the hand is not in contact with an object or a surface, and the joints have passive elastic elements, the following equation can be used to describe hand dynamics

$$\mathbf{B}(\dot{\mathbf{q}})\ddot{\mathbf{q}} + \mathbf{n}(\mathbf{q}, \dot{\mathbf{q}}) = \boldsymbol{\tau}_a + \boldsymbol{\tau}_f(\mathbf{q}, \dot{\mathbf{q}}) + \boldsymbol{\tau}_g(\mathbf{q}) + \boldsymbol{\tau}_p(\Delta\mathbf{q}), \tag{4.4}$$

where $\boldsymbol{\tau}_p(\Delta\mathbf{q})$ expresses the torque generated by the deformation of elastic elements, that can be evaluated as a function of joint deformation $\Delta\mathbf{q}$, evaluated w.r.t. a reference (rest) position of the hand $\mathbf{q}_0$, i.e., $\Delta\mathbf{q} = \mathbf{q} - \mathbf{q}_0$, $\boldsymbol{\tau}_f(\mathbf{q}, \dot{\mathbf{q}})$ expresses the equivalent joint torques due to frictional effects, $\boldsymbol{\tau}_g(\mathbf{q})$ models the gravitational terms, $\mathbf{B}(\dot{\mathbf{q}})$ represents the inertia matrix, and $\mathbf{n}(\mathbf{q}, \dot{\mathbf{q}})$ contains centrifugal and Coriolis terms. In this model we consider applications in which the velocities are limited, so that we can neglect the inertial and viscous terms. Furthermore, we assume that the torque generated by the deformation of elastic elements is higher than the one due to gravitational effects. The friction between tendons and finger parts could give a not negligible contribution, however its prediction depends on several parameters that cannot be easily considered (e.g., surface and material properties) and has not been included in the model for the sake of simplicity. With the above simplifications the dynamics of the system can be reduced to

$$\boldsymbol{\tau}_a + \boldsymbol{\tau}_p(\Delta \mathbf{q}) = 0 \tag{4.5}$$

Also this expression can be linearised if we consider a variation of the loading condition with respect to an initial reference configuration,

$$\delta \boldsymbol{\tau}_a = \mathbf{K}_q(\mathbf{q})\delta \mathbf{q}. \tag{4.6}$$

The linearisation leads to the introduction of the joint stiffness matrix $\mathbf{K}_q \in \Re^{n_q \times n_q}$, symmetric and positive definite, that can be evaluated, at a given configuration $\mathbf{q}$, as

$$\mathbf{K}_q = -\frac{\partial \boldsymbol{\tau}_p}{\partial \mathbf{q}}.$$

In the more general case, stiffness matrix $\mathbf{K}_q$ depends on joint configuration.

The problem that we want to solve is: how can we design hand joint stiffness $\mathbf{K}_q$ so that, when applying a certain force profile $\mathbf{f}_r(t)$ to the tendons, the fingertips of the hand follow a certain trajectory $\mathbf{p}_r(t)$? Fingertip trajectories could be defined for instance to perform a certain specific task, or according to mapping procedures as those described in [52], aimed at reproducing postural synergies identified for the human hand on robotic hands with dissimilar kinematic structures. We solved the problem by sampling the trajectory in a series of elementary steps and solving for each step the linearised equilibrium equation. The first step consists in solving the inverse kinematics for each finger, finding, for each time sample $t_k$ ($k$ indicates the generic time sample) the joint configuration vector $\mathbf{q}_d(t_k)$ corresponding to the desired fingertip configuration $\mathbf{p}_d(t_k)$. The variation of hand configuration corresponding to the sample $k$ is given by $\delta \mathbf{q}_k = \mathbf{q}_d(t_k) - \mathbf{q}_d(t_{k-1})$. This variation is produced by applying to the tendons a force variation $\delta \mathbf{f}_k = \mathbf{f}_r(t_k) - \mathbf{f}_\mathbf{r}(t_{k-1})$.

The following step consists in evaluating the stiffness matrix that allows to replicate such an elementary motion. Assuming that the $\mathbf{K}_q$ matrix is diagonal (i.e., the joints are independent), the problem can be solved straightforwardly: Eq. (4.4) can be rewritten, in this case, as

$$\delta \boldsymbol{\tau}_k = \mathbf{Q}_k \mathbf{k}_{q_k}, \tag{4.7}$$

where $\mathbf{Q}_k \in \Re^{n_q \times n_q}$ is defined as $\mathbf{Q} = \text{diag}(\delta \mathbf{q}_k)$, while $\mathbf{k}_{q_k} \in \Re^{n_q}$ is a vector collecting joint stiffness values at sample $k$. Taking into account Eq. (4.3), the system can be solved as follows

$$\mathbf{k}_{q_k} = \mathbf{Q}_k^{-1} \mathbf{T}_k^{\mathrm{T}} \delta \mathbf{f}_k. \tag{4.8}$$

where $\mathbf{T}_k$ is the tendon Jacobian matrix evaluated for the joint configuration vector $\mathbf{q}_k$. The solution of the linear system in Eq. (4.8) is a vector containing hand joint stiffness values at sample $k$ that allow to obtain a configuration variation $\delta \mathbf{q}_k$ of the hand when the tendons are pulled with a force variation $\delta \mathbf{f}_k$. It is worth to observe that stiffness values depend on both joint displacement $\mathbf{q}$ and on tendon force $f_r$. If we change the overall applied force to $\alpha \mathbf{f}_r$, where $\alpha > 0$ is a generic scaling factor, the same joint configuration can be followed by scaling the stiffness matrix with the same scaling factor, i.e., with a stiffness matrix $\alpha \mathbf{K}_q$. In other words, the shape of fingertip trajectories does not depends on $\mathbf{K}_q$ absolute values, but rather on their relative ratios. Normalizing vector $\mathbf{k}_q$ we obtain a base for the subspace of possible stiffness combination that can be used to track a desired trajectory. The values of $\mathbf{K}_q$ are defined on the basis of material properties and geometry constraints in the design of hand finger, as we will better detail in Sect. III. The procedure described so far leads, in the general case, to a stiffness matrix that depends on hand configuration, i.e., $\mathbf{K}_q = \mathbf{K}_q(\mathbf{q})$. Hand joints design to obtain a desired stiffness value may be not simple if the dependency of $\mathbf{K}_q$ on $\mathbf{q}$ is significant. In the following, we will show an application in which we obtain stiffness values that slightly depends on hand configuration.

### 4.3.2 *Modular Element Modeling*

Let us assume that the generic finger, indicated with index $j$ is realized by connecting $n_{q_j}$ modules with the same geometric dimensions, but with different stiffness values. Let us assume, for the sake of simplicity, that each finger is actuated by a single tendon, let us indicate with $t_j$ its length variation. The analysis can be easily extended to hands actuated with multiple tendons. Let us indicate with $l$ the length of the elastic element that constitute the joint, and with $h$ the distance between the elastic element bending axis and the tendon when the joint is in its rest position. According to the modular approach proposed in this chapter, we assume that this value is the same for all the joints, however the analysis can be easily extended to hands with different joints.

Let us then consider a generic joint $i$ on the finger $j$ of the hand. The main geometrical parameters adopted in this section are summarized in Fig. 4.5. When the $j$th tendon is pulled the elastic elements in the joints bend: let us suppose that the joint $i$ is rotated by an angle $q_i = \theta_i$, the corresponding variation of tendon length in the joint is indicated with $\Delta l_i$. We suppose that the bending elastic element assumes a circular arc shape, with radius $r_i = \frac{l}{q_i}$, and we indicate with $a_i$ the arc chord, that can be evaluated as $a_i = 2r_i \sin \frac{q_i}{2}$. The corresponding variation of tendon length can be evaluated as

$$\Delta l_i = l_i - \frac{a_i}{l}(r_i - h) = l - 2r_i \sin \frac{q_i}{2}\left(\frac{r_i - h}{r_i}\right) \tag{4.9}$$

The overall tendon displacement can be expressed as the sum of all the variations over finger joints, i.e., $t_j = \sum_{i=1}^{n_{qJ}} \Delta l_i$. We can observe that the relationship $t_j = t_j(\mathbf{q})$

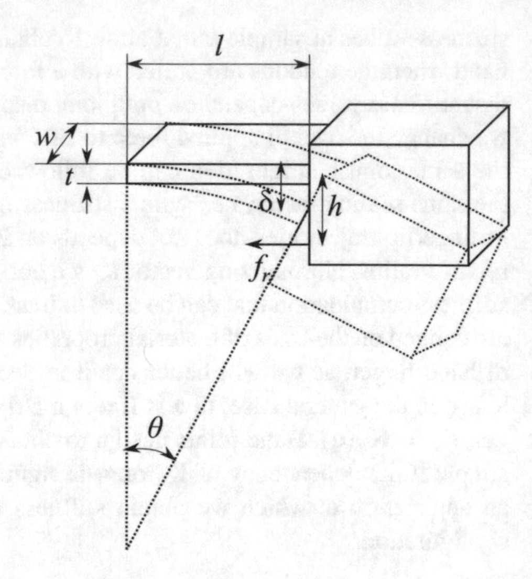


**Fig. 4.5** Main geometrical parameters of the modular passive joint

is nonlinear. The $(j, i)$th element of the Jacobian matrix $\mathbf{T}$ introduced in Eq. (4.2) can be evaluated as

$$T_{j,i} = \frac{\partial t_j}{\partial q_i} = -\cos\left(\frac{q_i}{2}\right)\left(\frac{l}{q_i} - h\right) + 2\frac{l_i}{q_i^2}\sin\left(\frac{q_i}{2}\right) \tag{4.10}$$

For small values of $q_i$, we can approximate $(\sin\frac{q_i}{2}) \approx \frac{q_i}{2}$, and we have

$$\Delta l_i \approx h\theta_i, \quad T_{j,i} \approx h. \tag{4.11}$$

For a $h/l$ ratio of about 0.2, i.e. the value that we adopted in the application that we will present in the following sections, the approximations in Eq. (4.11) lead to an error on $T_{i,j}$ value within 5% for $q_i < 0.3$ rad.

### 4.3.3 Passive Joint Design

In this subsection, we focus on hardware realization of soft joints with a given stiffness. Our approach is inspired to the work presented in [116] for soft grippers composed of flexible and stiff parts. The model takes into consideration the geometric and material properties of the robotic device. The realization of soft gripper using closed-top molding techniques is further detailed in an open source platform named Soft Robotics Toolkit [117]. Differently from that work, we exploited mechanical properties of flexible materials and their dependency on fabrication methods so to obtain different stiffness values of flexible parts through the variation of the percentage of infill density, that can be regulated during the 3D printing phase.

In this study, we furthermore follow a modular approach in the design: the basic idea is to realize modules composed of a flexible passive joint, in which the stiffness can be varied by properly regulating printing parameters, and a rigid link, which can be combined together to build underactuated soft robotic fingers. Using a sufficiently high number of modules, the deflection of each joint while the finger is bending is rather low, so that, for each joint, we can use the simplified linear results of beam theory. For higher deflections, a more complete bending model, as the one described in [118] should be adopted. When the actuator applies a force $f_{rj}$ through the $j$th tendon, it produces on the joint $i$ a torque $\tau_i = f_{rj}h$. We denote with $\delta_i$ the deflection of the elastic element realising the joint, that, according to beam theory, can be evaluated as

$$\delta_i = \frac{-f_{rj}hl^2}{2E_iI}, \tag{4.12}$$

where $E_i$ is the Young's modulus of the material, $I_i$ is the second moment of area, and $l_i$ is the length of the module elastic part. The corresponding joint rotation angle can be evaluated as

$$\theta_i = \frac{-f_{rj}hl_i}{E_iI_i}. \tag{4.13}$$

For the sake of simplicity we neglect the bending properties of the joint in the lateral and torsional direction. This simplification will be further validated through simulations in Sect. III.

The rotational stiffness of the joint can be then evaluated as

$$k_i = \frac{E_iI}{l}, \tag{4.14}$$

Assuming both parts of the module as filled rectangular shape whose centroid is located at the origin, the second moment of area is given by

$$I_i = \frac{w_it_i^3}{12}. \tag{4.15}$$

Referring to Eq. (4.14), we can observe that different joints stiffness can be achieved by changing either the geometric or material parameters of the modules. Since we aim at realising a modular structure, we assume that the geometric parameters are the same for each module, i.e. $I_i = I, l_i = l, h_i = h$. On the other hand, we can tune the material parameters in order to vary the modules stiffness.

### 4.3.4 Actuator Force Evaluation

In the stiffness design process, it is also important to evaluate the actuation force that is needed to bend the finger. The force applied by the actuator through the tendon produces a moment about the flexible part of the finger. The resultant behaviour can be approximated by a simplified cantilever beam model. The external forces acting on the finger are the force applied from the actuator and the interaction force which is represented in this model as a force applied on the fingertip $F_{tip}$. We can study the deflection effect generated by the two forces separately by using the superposition principle. Let $\delta_a$ be the deflection due to the actuator applied force and $\delta_r$ be the deflection due to the reaction force

$$\delta_a = \frac{f_r h l^2}{2EI}, \delta_r = \frac{F_{tip} l^3}{3EI}. \tag{4.16}$$

The sum of both deflections can be equated to zero and the resultant equation is solved for $F_{tip}$

$$F_{tip} = \frac{3 f_r h}{2l}. \tag{4.17}$$

Note that the terms $E$ and $I$ are cancelled out of the equation and as a result we do not need to consider the interaction between the alternating stiff and flexible parts of the modules to obtain the overall load at the fingertip. The model can be extended to any number of modules, as

$$F_{tip} = \frac{3 f_r h}{2 \sum_{i=1}^{2N} l_i}. \tag{4.18}$$

Furthermore, we can consider the parasitic capstan effects that takes place between the cable and stiff parts as the robotic finger transforms to a curve shape. As the modules lose the colinearity during their motion, the cable imparts a reaction force that resists further actuation. We can include the capstan effect by considering the angle between subsequent stiff parts of modules ($\theta_{i-1}$-$\theta_i$). The tendon does not pass through the flexible parts, so we can only consider the parasitic capstan effects on stiff parts. Thus, the fingertip force can be modified as

$$F_{tip} = \frac{3 f_r h}{2 \sum_{i=1}^{2N} l_i} \prod_{i=2,even}^{2n-2} e^{\mu(\theta_{i-1} - \theta_i)},$$

where $\mu$ is the friction coefficient.

### 4.3.5 Finger Module Stiffness Evaluation

Looking at the interesting additive manufacturing techniques that nowadays are rapidly increasing and offering interesting new opportunities, we analyzed the possibility of tuning finger joint stiffness values through exploiting the potentialities of 3D printing fabrication methods. In particular, choosing a material as the Thermoplastic

**Table 4.2** TPU mechanical properties as a function of 3D printing infill density percentage, from [119]

| Infill density $\rho\%$ | E (MPa) |
|---|---|
| 10 | 1.07 |
| 30 | 1.38 |
| 50 | 2.07 |
| 70 | 6.53 |
| 90 | 9.45 |
| 100 | 10.50 |

Polyurethane for realizing the flexible parts, we can get different stiffness values, while maintaining the same geometric shape, by regulating the percentage of infill density. This parameter affects primarily material density, but also its mechanical properties. As an example, Table 4.2 summarizes the variation of Young's modulus $E$ of the Thermoplastic Polyurethane as a function of the infill percentage density $\rho$ [119].

For a given geometric structure of the joint, the corresponding stiffness value is therefore directly proportional to the infill density

$$k_i = f(\rho_i). \tag{4.19}$$

As an example, we evaluated with the Finite Element Method (FEM) based software COMSOL the stiffness of a modular element varying material properties. The dimensions of the elements are: length $l = 13$ mm, thickness $t = 2.5$ mm, width $w = 21$ mm. Figure 4.6 and Table 4.3 summarizes the main obtained results, in terms of stress and deformation of each module, and joint stiffness for different values of infill density percentage. Another parameter that we need to consider is the passive flexibility of the finger in the lateral direction and with respect to a torsional load, that cannot be controlled by actuators, and that could affect system functionality. We therefore evaluated also the stiffness of each module in the lateral direction, $k_l$, and the torsional stiffness, $k_t$. We reported the main results of this analysis in Fig. 4.6 and Table 4.3. We can observe that, for the given geometry, the lateral stiffness is much higher than the bending one, due to the module geometry, while the torsional one is closer. A finger composed of simple modules as those considered in this analysis may therefore have an excessive torsional compliance, that however could be potentially compensated by modifying the design of the rigid part of the model.

### 4.3.6 Simulations

To let the Soft-SixthFinger be able to grasp a wide range of objects, it is important to define a suitable finger flexion trajectory. To this aim, we took inspiration from the motion of the human hand. In particular, we used the mapping algorithm originally

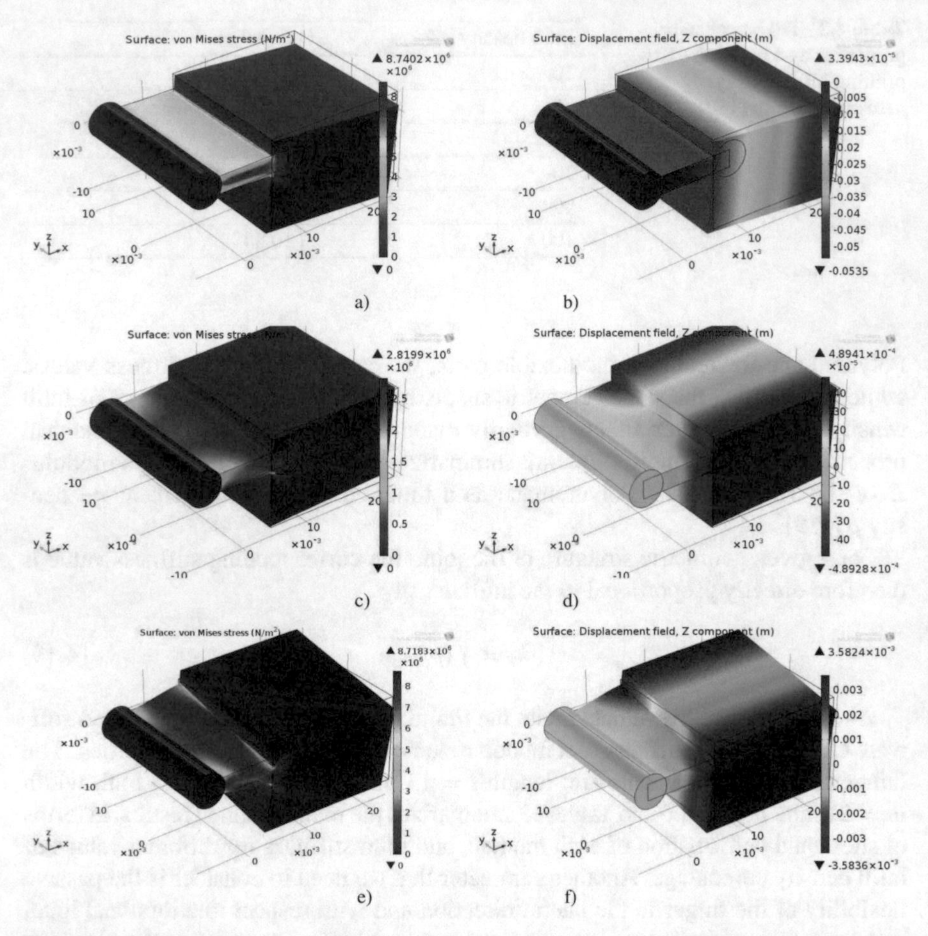


**Fig. 4.6** FEM model and stress/deformation analysis of the passive joint. In all the simulations, the left part of the flexible link was constrained to be fixed, while a load was applied to the rigid module connected to the right. Different loading conditions were applied to estimate the equivalent stiffness in different directions. The diagrams reported here summarizes the results obtained for the 100% of infill density. **a**, **b** bending in the principal direction, a 5 N load was applied to the free boundary of the rigid module in the $z$ direction. **a** stress distribution, **b** module deformation. **c**, **d** bending in the lateral direction, a 5 N load was applied to the free boundary of the rigid module in the $y$ direction. **c** stress distribution, **d** module deformation. **e**, **f** torsion, a ±5 N load was applied to the free lateral boundaries, so to produce an equivalent torsional moment of about 100 N mm with respect to the $x$ direction. **e** stress distribution, **f** module deformation

proposed in [23, 52] to transfer the motion of the human hand onto robotic hands with dissimilar kinematics. The mapping method allows to replicate on the Soft-SixthFinger, the effects in terms of motions and deformations that a human reference hand would perform on a virtual object. This allows to work directly on the task space avoiding a specific projection between different kinematics. Details of the mapping are not reported here for the sake of brevity, interested reader can refer to [23, 52]. We

**Table 4.3** Results from FEM analysis of the passive joint, stiffness value as a function of infill density percentage of the material, for a given geometry

| Infill density $\rho\%$ | $k(\rho)$ Nmm/rad | $k_l$ Nmm/rad | $k_t$ Nmm/rad |
|---|---|---|---|
| 10 | 4.6930 | 331.3 | 24.9 |
| 30 | 6.0526 | 427.2 | 32.1 |
| 50 | 9.0789 | 640.8 | 48.1 |
| 70 | 28.6403 | 2021.6 | 151.7 |
| 90 | 41.4474 | 2925.6 | 219.7 |
| 100 | 46.052 | 3807.9 | 285.7 |

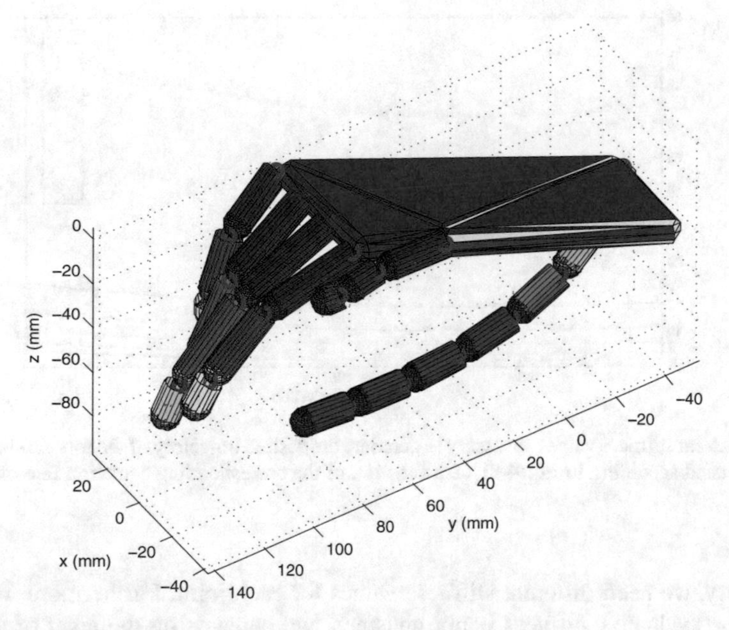


**Fig. 4.7** The SynGrasp model of the human hand wearing the Soft-SixthFinger used to compute the joint stiffness values

considered a model of a human hand augmented with a model of the robotic finger, see Fig. 4.7. We simulated using the Matlab SynGrasp toolbox [120] the motion of the human hand according to the first synergy as defined in [121]. We then computed the trajectory of the robotic fingertip using the mapping algorithm. This target trajectory in the Cartesian space is the result of the mapping of the first human hand synergy onto the robotic device.

Once a target trajectory is defined, we take advantage of the stiffness design presented in Sect. 4.3.3 to simulate a tendon driven underactuated finger able to track the desired fingertip trajectory. Considering polyurethane as material for the flexible joints, we computed the values for $\mathbf{k}_q$ reported in Table 4.3. To obtain the desired

**Table 4.4** Joint stiffness values computed by model for TPU material and corresponding infill density percentage

| Stiffness values (Nmm/rad) | Infill density $\rho\%$ |
|---|---|
| 6.58 | 14.3 |
| 7.67 | 16.7 |
| 9.2 | 20 |
| 11.5 | 25 |
| 15.3 | 33.4 |
| 23.02 | 50 |
| 46.05 | 100 |

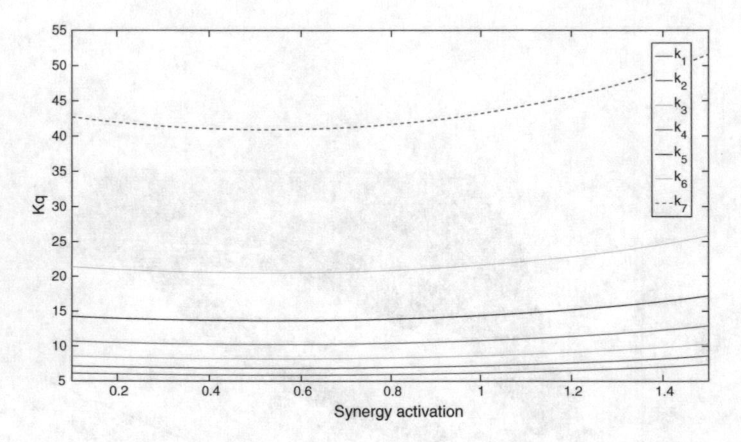


**Fig. 4.8** Joint stiffness values necessary to generate the desired trajectory of the robotic sixth finger tip, evaluated according to Eq. (4.8), as a function of the corresponding actuation rate of the first synergy

trajectory, we need different stiffness values for each joint. Furthermore, for each joint, the evaluated stiffness is not constant, and depends on the level of synergy actuation, with a nonlinear behavior. Since: (1) it is not practically feasible to design the stiffness of each joint so that it follows the evaluated relation, (2) the stiffness variation for each joint for the proposed trajectory is low and (3) the trajectory shape depends on the ratios between the different stiffness values, rather than on their actual value, we then evaluated the trajectory obtained if each joint stiffness is assumed constant. The stiffness values and corresponding percentage infill density used are reported in Table 4.4. The range of stiffness values for the joints was decided according to the properties of the material used for the soft joints, i.e., a thermoplastic polyurethane (Lulzbot, USA) called "NinjaFlex". We selected the range of values that were possible to replicate using that material and a 3D printer. We can therefore approximate the desired trajectory by realizing joints with different, but constant, stiffness values (Fig. 4.8).

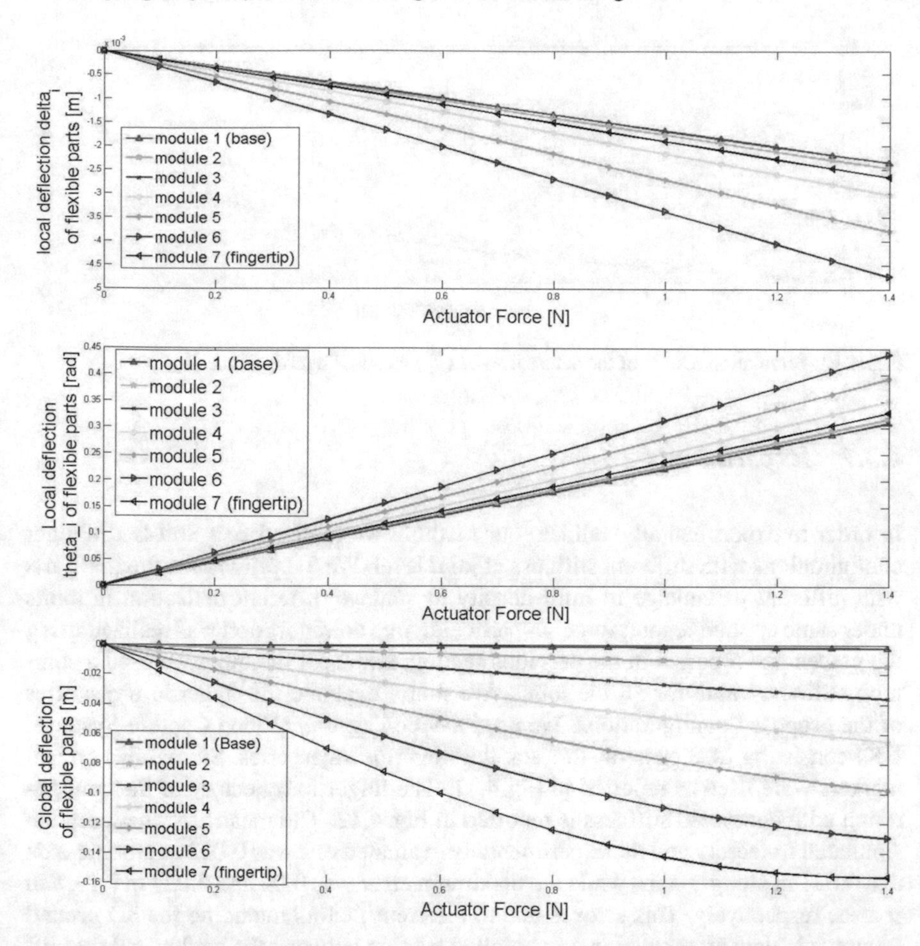


**Fig. 4.9** The simulation results: local bending $\delta_i$ in each module versus actuator force (top), local theta $\theta_i$ for each module versus actuator force (middle) and overall bending of each module with respect to the base module versus actuator force (bottom), are shown

Once we have the target trajectory and the relative values of joint stiffness, we can use the proposed numerical model to study the kinematics of the tendon-driven Soft-SixthFinger and to simulate its fingertip trajectory. We simulated the model with the infill density percentage reported.

The simulation results corresponding to the computed stiffness values are plotted in Figs. 4.9 and 4.10. In particular, Fig. 4.9 shows the bending deflection ($\delta_i$) and the angle ($\theta_i$) for each module with respect to its local coordinates and the overall deflection of each joint with respect to global coordinates defined at the base of finger. On the other hand, Fig. 4.10 reports the plot of the actuator applied force to fingertip force.

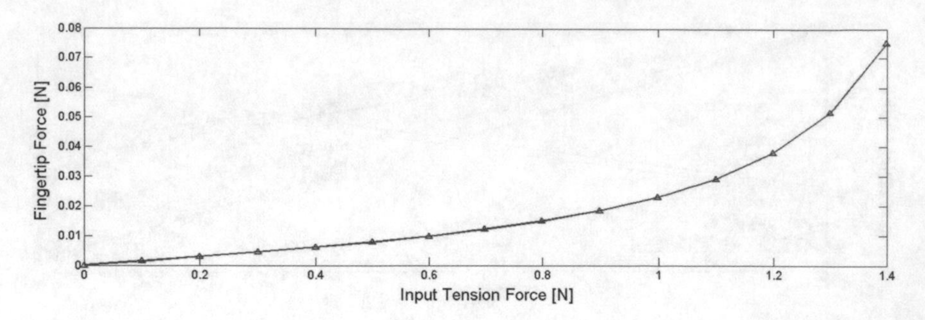


**Fig. 4.10** Simulation results of the actuator force ($f_r$) versus Fingertip Force ($F_{tip}$)

## *4.3.7 Experiments*

In order to experimentally validate our method, we realized two Soft-SixthFinger configurations with different stiffness at joint level. We 3D printed the flexible parts with different percentage of infill density to achieve different deflection in joints under same applied tendon force. In particular, one configuration was realized using the percentage reported in the previous section, while for the other we used a common stiffness value for all the joints. We firstly evaluate the fingertip trajectories of the proposed configurations. We used a Vicon system (Vicon Capture Systems, UK) consisting of 8 cameras to track the fingertip trajectories. Six passive optical markers were used as reported in Fig. 4.11. The fingertip trajectory of the configuration with computed stiffness is reported in Fig. 4.12. The mean error between the simulated trajectory and the experimentally evaluated one was 0.023 m along $x$-axis and 0.021 m along $y$-axis, while the maximum error was 0.04 and 0.027 m for $x$ and $y$ axes, respectively. This error is due to different factors including the 3D printed accuracy in density regulation, unmodelled friction between the tendon and the stiff part of the modules and small finger fluctuation during the flexion motion.

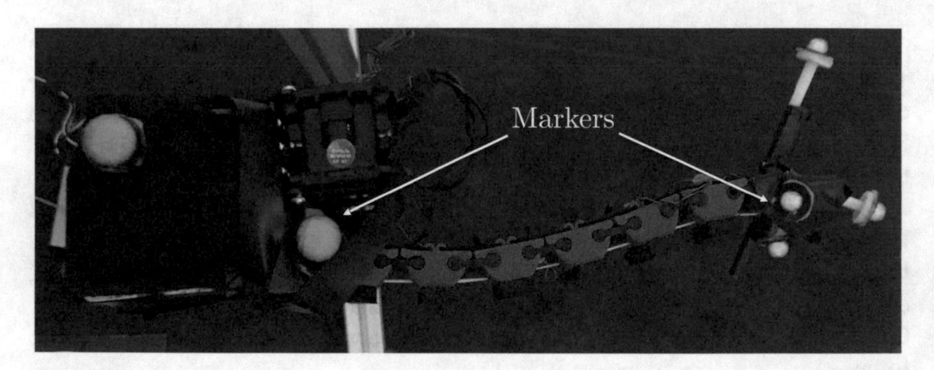


**Fig. 4.11** The setup for fingertip tracking using the Vicon system

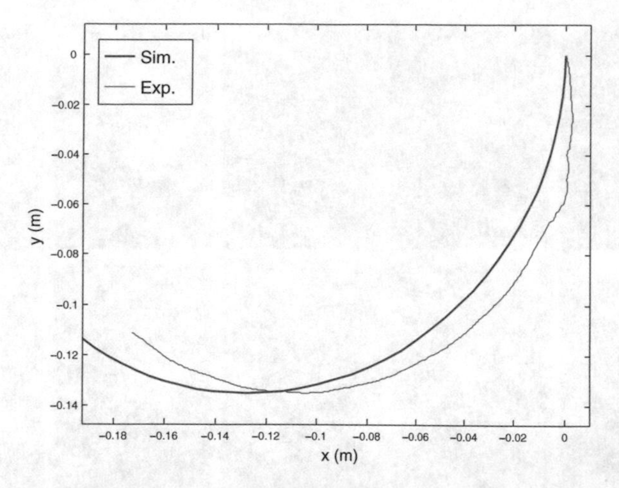


**Fig. 4.12** Trajectory of the fingertip of the Soft-SixthFinger simulated (*red*) and experimentally evaluated (*blue*). Simulation are obtained using the model presented in Sect. II. Experimental data are obtained using an optical tracking system

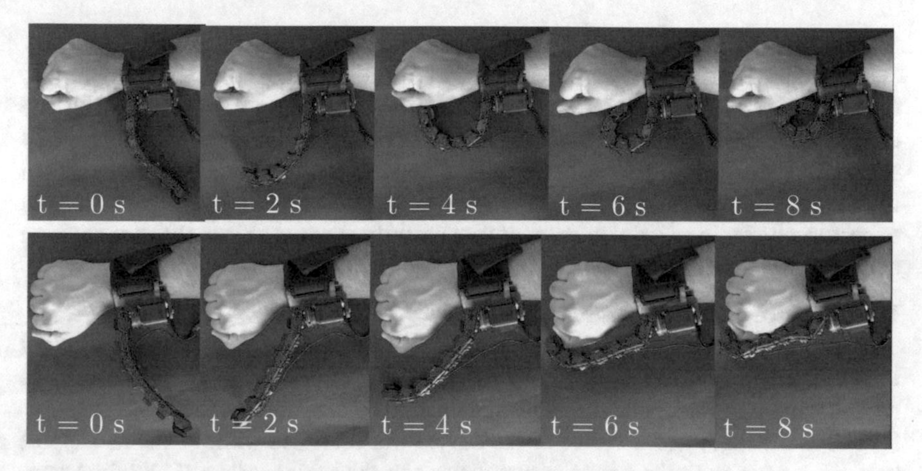


**Fig. 4.13** Trajectories of the Soft-SixthFinger using different stiffness at joint level. On top the joint stiffness is computed using the design method introduced in this study. On the bottom, the trajectory obtained considering all the joints with the same stiffness value

In order to evaluate how the joints' stiffness regulate the trajectory of the finger and, consequently, how the flexion trajectory influence the device adaptability to different shapes of the grasped objects we performed a qualitative experiment. A healthy subject was asked to wear the devices on its right arm, while simulating the paretic hand. Figure 4.13 shows the snapshots of the two configurations of Soft-SixthFinger with different joints stiffness. The top configuration has the stiffness in each joint computed using the design method introduced in this chapter. The the bottom one shows the configuration with the same stiffness in the all the joints.

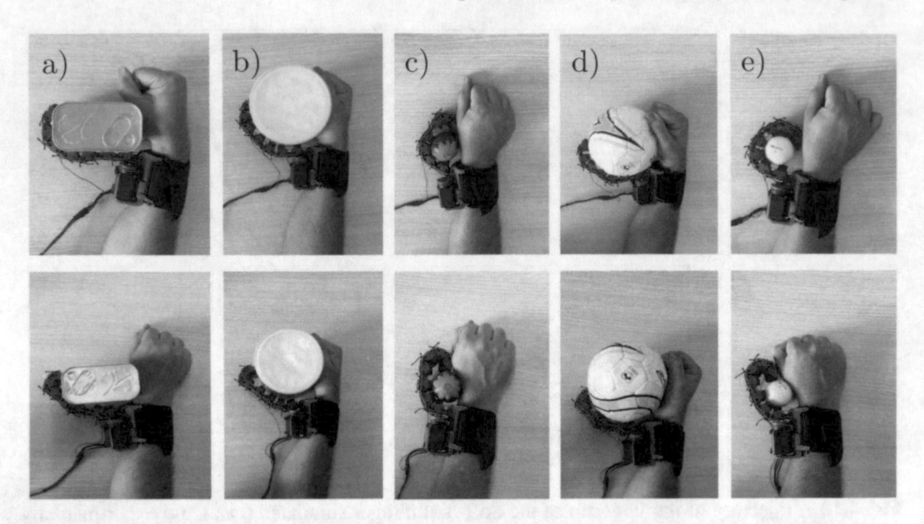


**Fig. 4.14** The soft-SixthFinger with two different trajectories grasping various objects with different shapes and sizes: **a** meat can, **b** coffee can, **c** strawberry, **d** soft ball **e** golf ball

Furthermore, to test the grasping ability of the device with the realized configurations, we used a subset of the objects included in the YCB grasping toolkit [122]. The objects were selected so to consider different object sizes and shapes. Figure 4.14 shows the snapshots with some of the grasped objects with both configurations shown in Fig. 4.13. It was noticed that the configuration with the computed stiffness was able to perform better in terms of number of successfully grasped object, shape adaptation and grasp performance.

**Performance Characterization**

The device performances were evaluated through a subset of the tests proposed in [123]. In particular, we measured the maximum fingertip force, the maximum payload and maximum horizontal grasp resistive force.

The maximum fingertip force of the device was recorded while fixing its support base on a table with the finger perpendicular to the table surface. The initial configuration of the finger was fully extended and it was commanded to close at the maximum torque. The hook of a dynamometer (Vernier, USA) was rigidly coupled with the fingertip of the device so that force could be measured in vertical direction as shown in Fig. 4.15a. The constant applied force value at fingertip is presented in Table 4.5. The maximum horizontal grasp resistive force was measured by grasping an object (diameter = 65 mm, weight = 400 g) with the robotic device and the arm. The object was slowly pulled horizontally by using the hook of the dynamometer, see Fig. 4.15c. It was measured that the grasp remained stable till 13 N. To check the maximum payload, an operator wore the grasp compensatory robotic device. The operator's arm was stabilized on a table while grasping a cylindrical object (diameter = 65 mm, weight = 400 g) with the aid of the robotic tool at its maximum actuator's torque. The grasped object was slowly pushed down using the dynamometer's

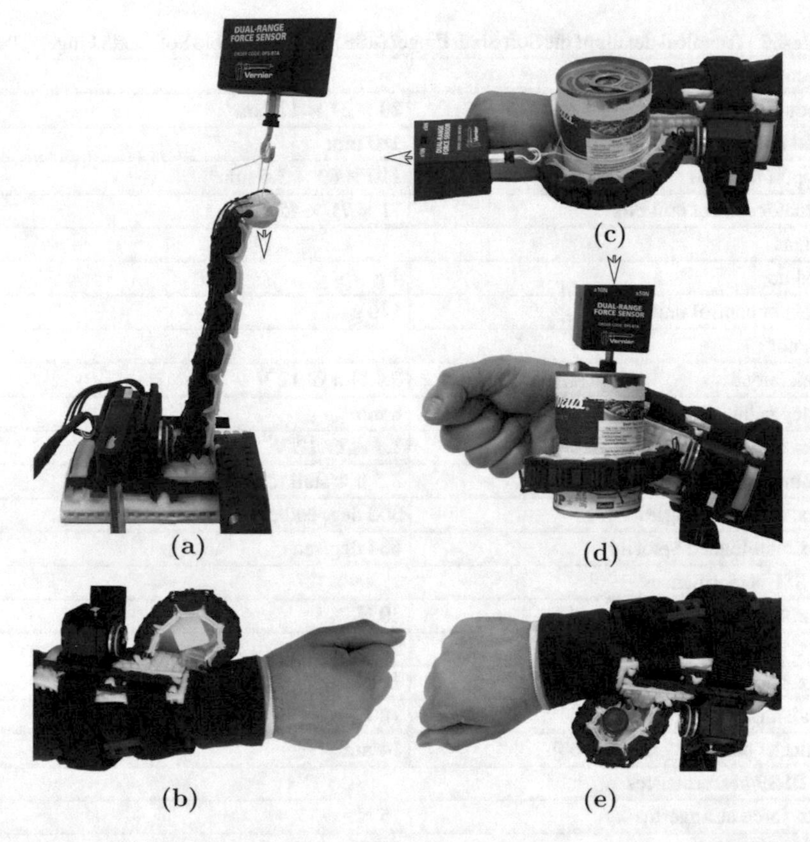


**Fig. 4.15** Experimental procedure to measure the performance characteristics of the soft sixth finger. The figure **a**, **c**, **d** show the experimental setup for maximum fingertip force, horizontal grasp resistive force and maximum payload, respectively. The arrow in figure show the direction of applied force. Figure **b** and **e** show the minimum size graspable objects

bumper as shown in Fig. 4.15d. The maximum pushing force was recorded when the object started to slip. The maximum payload of the device (see, Table 4.5) is the sum of grasped object's load and the one due to the pushing force. Figure 4.15b, e show the minimum size graspable objects. The diameter of the smallest graspable object is reported in Table 4.5.

### 4.3.8 The Double Soft Sixth Finger

Although the soft sixth finger can be used to grasp and stabilize a large set of objects, having a single finger in opposition to the patient arm can result in a limitation in tasks requiring a high payload. We designed the double soft sixth finger to deal with these

**Table 4.5** Technical details of the Soft Sixth Finger (SSF) and the Double Soft Sixth Finger (DSSF)

| Dimensions | |
|---|---|
| Module | $20 \times 31 \times 12$ mm$^3$ |
| Total length of finger (on arm) | 180 mm |
| Support base | $110 \times 63 \times 3.5$ mm$^3$ |
| Actuator control unit box | $71 \times 71 \times 45$ mm$^3$ |
| Weights | |
| Module | 4 g |
| Actuator control unit box | 146 g |
| Actuator | |
| Max. torque | 3.1 Nm @ 12 V |
| Pulley radius | 8 mm |
| Max. current | 1.4 A @ 12 V |
| Continuous operating time | 3.5 h @stall torque |
| Max. operating angles | 300 deg, endless turn |
| Max. non-loaded velocity | 684 deg/sec |
| The SSF performances | |
| Max. force at fingertip | 40 N |
| Max. payload | 2.4 kg |
| Max. horizontal resistive force | 13 N @ dia = 65 mm |
| Total: finger + support base | 180 g |
| Diameter smallest graspable obj. | 14 mm |
| The DSSF performances | |
| Max. force at fingertip | 15 N |
| Max. payload | 4.87 kg |
| Max. horizontal resistive force | 26 N @ dia = 65 mm |
| Total: finger + support base | 230 g |
| Diameter smallest graspable obj. | 17 mm |

particular situations. The double soft sixth finger shares with the soft sixth finger the same principle design guidelines related to wearability, modularity, symmetrical structure and underactuation. It is composed by two parts: a support base that allows the finger to be worn at the patient forearm and two fingers. The angle between the two finger has been selected so to maximize the distance between the two fingertips, while not exceeding the support base size. The exploded view of unit module and complete double soft sixth finger is shown in Fig. 4.16. Two tendon wires (one for each flexible finger) and a single actuator control the motion of the device. One end of each tendon wire is fixed to each fingertip, while the other ends of both tendon wires are attached to a single pulley mounted on the shaft of the actuator (MX-28T). When the motor rotates, both tendon wires are wound on the pulley and fingers are flexed to grasp the object. As the motor is rotated in opposite direction, the elastic parts

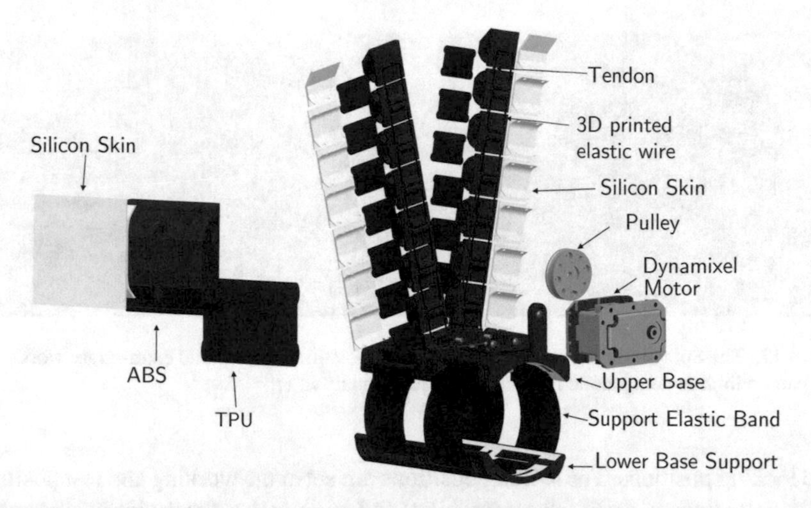


**Fig. 4.16** The CAD exploded view of the double soft sixth finger. On the left unit module with single tendon. On the right, the exploded view of complete double soft sixth finger

in the joints restore the finger to its extended configuration. We performed similar evaluation for double soft sixth finger to quantify its payload, maximum fingertip force and horizontal grasp resistive force. The results of the experiments are shown in Table 4.5. The approach of building compensatory robotic device using two robotic fingers improved the payload and horizontal grasp resistive force which in turn can handle relatively heavier objects. The shape adaptation to the grasped object was also confirmed by using the objects in the YCB toolkit.

### *4.3.9 Wearability and Device Positioning at Forearm*

The actuator's controller and battery have been enclosed in a small box to be worn on the patient's belt. Only the device and its actuator are placed at the patient's forearm to keep minimum weight on it.

The device can be worn by the user without any assistance, just inserting their hand/arm between the two base parts. Then, using the velcro elastic band, the patient can tight the device at the forearm. The robotic devices can be wrapped up on the arm as bracelets to reduce the encumbrance when being not used as shown in Fig. 4.17. The patient can use his or her healthy hand to switch from the rest to the working position and viceversa. The switching between the two positions is achieved through a passive rotatable Dovetail triangular locking mechanism. The mechanism consists of two parts, namely A and B. Part A is embedded on the support base while part B is contained in the finger starting module. After coupling both parts together, a pin joint has been added in the center of both parts to allow only rotation while constraining the decoupling of both parts without unscrewing the pin joint. The mechanism has

**Fig. 4.17** The Soft-SixthFinger in rest and working position. The robotic extra-finger works with the paretic limb to compensate for hand grasp functionality

two locking positions. The locking positions are set at the working and rest positions of the extra fingers. Apart from wearability and ergonomics, the device position at the forearm plays an important role in the task performance. The location of the device depends on the patient conditions and on the residual mobility of the arm/hand. The compensatory robotic devices can be worn on the distal part of the forearm (near or on the wrist), so to obtain the grasp by opposing the device to the paretic hand. However, the distal position of the robotic finger may fail when the motor deficit is so advanced that a pathological synergism in flexion has taken place. In this case, the wrist and fingers are too much flexed not allowing successful grasping. When this pathological condition occurs, the extra-finger may be positioned more proximal at the forearm, so to let the grasp be achieved by the extra-fingers opposition to the radial part of the wrist. This flexibility in the positioning is achieved thanks to the symmetrical structure and the ergonomics of the support base. The support base of the fingers can be translated or rotated along the forearm to place the finger on a suitable orientation. These features enable the device to adapt the patient conditions and increases the versatility of the device. Moreover, the elastic straps along with velcro enable the devices to fit to different size of arms and facilitates the patients in wearing robotic device himself without any assistance.

## 4.4 Tests with Chronic Stroke Patients

We performed a series of experiments with five chronic stroke patients (four male, one female, age 40–62) to prove the effectiveness of the devices in grasp compensation. Written informed consent was obtained from all participants. The procedures were in accordance with the Declaration of Helsinki. We targeted ADL bi-manual tasks to evaluate if the compensatory robotic devices can assist the patients. In order to use the proposed devices, the subjects should have residual mobility of the arm. For being included in the experimental phase, patients had to score $\leq 2$ when their motor function was tested with the National Institute of Health Stroke Scale (NIHSS) [84], item 5 "paretic arm". Moreover, the patients had to show the following characteristics: normal consciousness (NIHSS, item 1a, 1b, 1c $= 0$), absence of conjugate eyes deviation (NIHSS, item 2 $= 0$), absence of complete hemianopia (NIHSS, item $3 \leq 1$), absence of ataxia (NIHSS, item 7 $= 0$), absence of completely sensory loss (NIHSS, item $8 \leq 1$), absence of aphasia (NIHSS, item 9 $= 0$), absence of profound extinction and inattention (NIHSS, item $11 \leq 1$).

The goal of the tests was to evaluate how quickly the patients could learn to use the devices and which device they preferred to fulfill a certain task. The patient were asked to select between the soft sixth finger and the double soft sixth finger to perform the proposed tasks. Patients wore the robotic device on their paretic limb, on the left for two subjects and on the right one for the other three. The rehabilitation team assisted the subjects during a training phase that lasted about one hour. During this phase, the optimal position of the device on the arm was evaluated according to the patient motor deficit. After the training phase, the subjects were asked to perform a list of bi-manual tasks with the aid of proposed devices. We proposed the patients three possible scenarios. The first included two different kitchen activities involving multiple bi-manual tasks. In the second, we tested the device with bi-manual tasks using tools. Finally, we tested the use of the robotic extra fingers to carry a shopping bag while walking.

### 4.4.1 Kitchen Scenario

Cooking in the kitchen involves a variety of bi-manual tasks and many of them are based on hold and manipulate techniques. The compensatory robotic devices can support the patients in performing such tasks even if one hand is paretic. Figures 4.18 and 4.19 show the snapshots of the tasks performed by the subjects to *prepare the breakfast* and *lunch*.

**Preparing Breakfast**

We asked the patients to simulate the activities of preparing coffee, putting jam on bread slice and peeling apple for breakfast.

*Task 1 "Opening the coffee pot"*: Hold firmly the base of coffee moka pot with the help of robotic device while use healthy hand to unscrew the upper part (see, Fig. 4.18a).

*Task 2 "Closing the coffee pot"*: Fill the filter with a mixture of coffee grounds. Grasp the base part with the device and close the pot again.

*Task 3 "Pouring coffee into cup"*: pour coffee using healthy hand while holding cup with the compensatory device (see Fig. 4.18b).

*Task 4 "Opening jam jar"*: grasp the jar with the device and non-functional arm while opening the cap with functional one.

*Task 5 "Spreading jam on bread"*: grasp the jam jar to take jam from it by holding the knife in healthy hand. Spread jam on a bread slice using functional hand(see, Fig. 4.18c).

**Fig. 4.18** Preparing breakfast with the help of grasping compensatory tool. **a** opening the Moka pot, **b** pouring coffee into cup, **c** spreading jam on bread, **d** peeling apple

*Task 6 "Peeling apple"*: grasp the apple with device and peel it using knife in the healthy hand(see, Fig. 4.18d).
All the patients choose the soft sixth finger to perform the task, since the single finger results more suitable while manipulating relatively light weight and smaller objects.

**Preparing Lunch**
*Task 1 "Opening tomato sauce jar"*: constrain the motion of the tomato jar with the device and paretic arm while the healthy hand unscrews its cap (see, Fig. 4.19a).
*Task 2 "Pouring"*: pouring the tomato sauce from its jar into a cooking pot.
*Task 3 "Opening tuna can"*: hold the tuna can with the device while functional hand open its cap (see, Fig. 4.19b).
*Task 4 "Opening beans can"*: constrain the motion of beans can with compensatory robotic device and paretic arm while the healthy hand open its cap.
*Task 5 "Pouring"*: Pouring the beans from its can into a cooking pot.

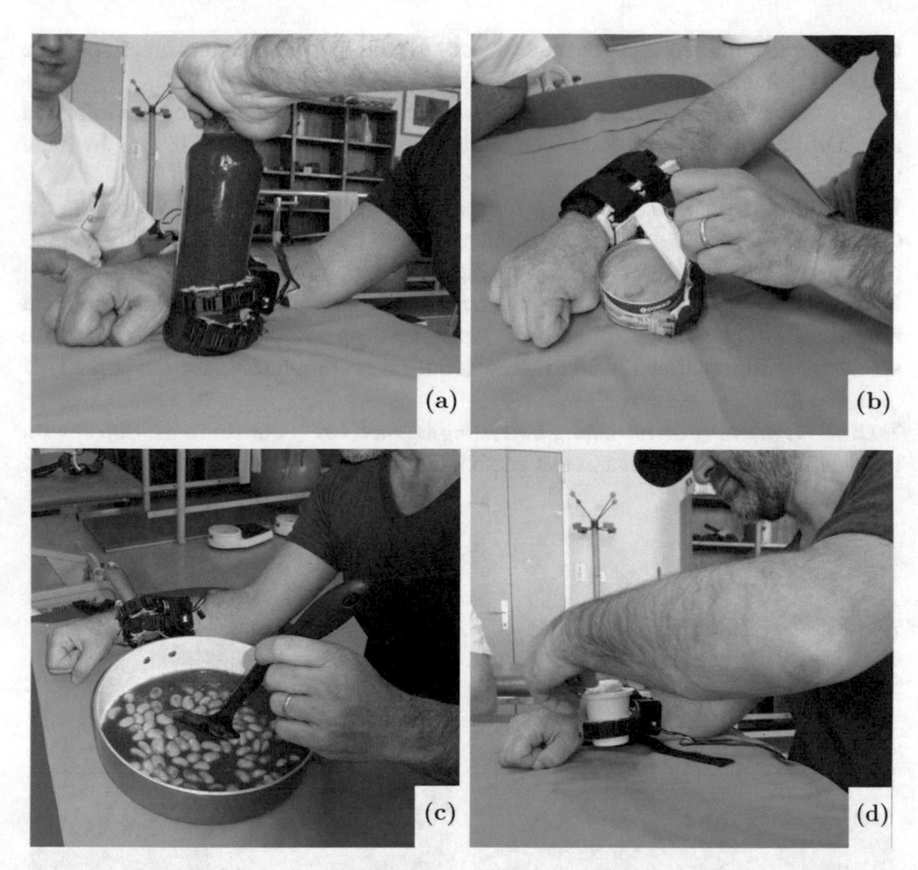


**Fig. 4.19** Preparing lunch with the help of grasping compensatory robotic device. **a** opening tomato sauce jar, **b** opening tuna can, **c** stirring food, **d** opening yogurt cup

*Task 6 "Stirring"*: hold firmly the cooking pot with the compensatory robotic device while functional hand stirrs the food in it (see, Fig. 4.19c).
*Task 7 "Opening yogurt cup"*: constrain the yogurt cup and remove its cover (Fig. 4.19d).

All patients decided to use the double soft sixth finger to perform tasks (a) and (c), while the soft sixth finger was selected to complete tasks (b) and (d). Snapshots of the execution of the tasks are reported in Fig. 4.19.

### 4.4.2 Tools Activities

The tools activities are another example of ADL where many tasks are based on hold and manipulate principle. The presence of the compensatory robotic device can help

the patient to complete such bi-manual tasks even if one hand is non-functional. We asked the patients to use the tools to perform the following tasks.
*Task 1: "Drilling in wood block"*: grasp the wood block with the device and impaired arm. Drill a hole in the wood block while using drilling machine in healthy hand (see, Fig. 4.20a).
*Task 2: "Removing nail using claw hammer"*: hold firmly the wood block with the compensatory robotic device and paretic arm. Use claw-hammer in healthy hand to pull the nail from the wood block (see, Fig. 4.20b).
*Task 3: "Inserting tapcon screw in the drilled hole"*: use device and non-functional arm to hold the object of drilled hole. Place the tapcon screw at hole position, use screw driver in functional hand to screw it until it is completely inserted in the hole (see, Fig. 4.20c).
*Task 4: "Tightening or loosening bolt using wrench key"*: constrain the object with the device and paretic arm and use wrench key in healthy hand to tight or loose the

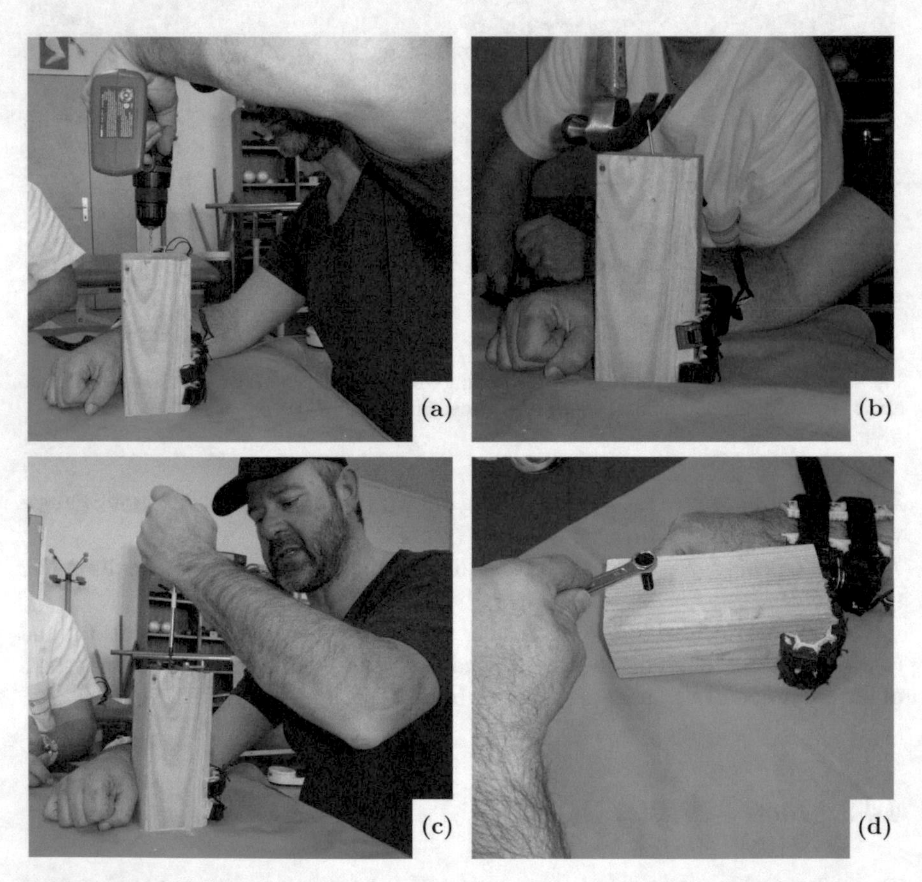


**Fig. 4.20** Using different tools in bi-manual tasks with the aid of robotic device. **a** drilling in wood block, **b** removing nail using claw hammer **c** inserting tapcon screw **d** tightening or loosening bolt

bolt (see, Fig. 4.20d). double soft sixth finger was used in all tasks to assure stable and firm grip on the wood block. All patients selected the double soft sixth finger to perform the tasks.

### 4.4.3 Active Hook

The last application proposed was to carry shopping bag with the compensatory robotic device while walking as shown in Fig. 4.21. The patients were successfully able to carry the bag using robotic devices on paretic arm. As expected, double soft

**Fig. 4.21** Carrying shopping bag with the help of grasping compensatory robotic device

sixth finger was able to carry heavier bag as compared to the soft sixth finger due to its higher payload.

### 4.4.4 *Questionnaires*

After the experiments, we asked the patients about the usefulness and possible concerns related to the compensatory robotic devices for performing ADL tasks. The patients were asked to fill the Usefulness-Satisfaction-and-Ease-of-use questionnaire (USE, [124]) that focuses on the experience of the system usage. This questionnaire uses a seven-point Likert rating scale. Mean and standard deviation (SD) of the questionnaire factors are presented in Table 4.6. Moreover, to evaluate the patient's satisfaction to the proposed compensatory devices and their features, we asked the patients to fill the first part of the QUEST 2.0 questionnaire [125]. The purpose of the QUEST questionnaire is to evaluate how satisfied patients are with the proposed assistive device. The mark ranges from "1 = not satisfied at all" to "5 = very satisfied". Mean and standard deviation (Mean (SD)) are reported in Table 4.7.

**Table 4.6** Questionnaire factors and relative marks. The mark ranges from "1 = strongly disagree" to "7 = strongly agree". Mean and standard deviation (Mean (SD)) are reported

| Questionnaire factors | Mean (SD) |
|---|---|
| Usefulness | 5.6(0.8) |
| Ease of use | 6.0(0.6) |
| Ease of learning | 6.5(0.8) |
| Satisfaction | 5.8(0.7) |

**Table 4.7** Quebec User Evaluation of Satisfaction with assistive Technology. The mark ranges from "1 = not satisfied at all" to "5 = very satisfied". Mean and standard deviation (Mean (SD)) are reported

| How satisfied are you with | Mean (SD) |
|---|---|
| The **dimensions** (size, height, length, width) of your assistive device? | 4.6(0.8) |
| The **weight** of your assistive device? | 4.0(0.6) |
| How **safe** and **secure** your assistive device is? | 4.5(0.8) |
| The **durability** (endurance, resistance to wear) of your assistive device? | 4.6(0.7) |
| How **easy** it is to **use** your assistive device? | 4.6(0.7) |
| How **comfortable** your assistive device is? | 4.2(0.7) |
| How **effective** your assistive device is (the degree to which your device meets your needs)? | 4.0(0.8) |

## 4.5 Results and Discussion

The ergonomics and functional requirements listed in Table 4.1 have been considered in the design and development of the robotic devices. Table 4.8 summarizes the devices achieved requirements.

The robotic extra fingers have been tested with different targets in order to demonstrate how the soft fingers can adapt to the shape of the object, producing a stable enveloping grasp. We tested the device with five chronic stroke patients in ADL. The proposed robotic devices successfully enabled the patients to complete all the presented bi-manual tasks. The experiments authenticated that the presented robotic devices can be an effective aid for chronic stroke patients to perform simple ADL tasks. The patients questionnaire feedback showed the effectiveness of the proposed compensatory robotic devices in assisting ADL tasks. The proposed robotic devices resulted to be an effective aid in completing the ADL bi-manual tasks. If compared to the old versions of the device, the soft sixth finger showed better performance due to the new actuation, the more stable support base and the increased friction at contact points. The realization of a new device, the double soft sixth finger, increased the potential use of compensatory devices in the ADL tasks, since it is able to realize a more stable grasp in relatively more payload demanding and pouring tasks. At this early stage of research we cannot determine which device is better to fulfill a certain task. The soft sixth finger resulted more easy to use when it is necessary to grasp small objects. It is also more wearable and portable with respect to the version with two fingers. On the other hand, the double soft sixth finger has an higher payload and can be used in more complicated manipulation tasks, such us pouring water from a bottle. This improvement on grasp stability come at the cost of an higher weight and a reduced wearability. We believe it is worth exploring both the solutions in different applications and different tasks and keep developing the two platforms in collaboration with the clinicians and the patients. Although the experiments with patients showed the effectiveness of the devices in the completion of some ADL, at the moment this approach has some limitations. Patients suffering from hemiplegia or hemiparesis can vary over a wide range, from mild weakness and loss of dexterity in the fingers to complete paralysis in the left or right side of the body. Although the proposed compensatory devices are able to compensate in terms of grasping, the use of the device require somc mobility in the impaired arm with even non-functional hand. Complex manipulation bi-manual tasks like tying shoelaces or buttoning, are too demanding and out of scope of both selected group of patients and current devices. However, many ADL tasks including those presented in the chapter, can successfully be completed with the aid of proposed robotic devices .

**Table 4.8** How the proposed robotic devices meet the ergonomics and functional requirements listed in Table 4.1

| Category | Requirement | Actual realization |
|---|---|---|
| Ergonomics | Extreme Wearability | The devices can be worn through support base and can be shaped into bracelet when being not used |
| | Portability | Portable complete system including power supply and actuator control unit. The wireless communication between eCap and robotic device |
| | Weight | Maximum 230 g at arm |
| | Ease of use | Ease in wearing is realized by eCap, support base and elastic straps. Easy working principle and few control inputs from user |
| Functional | Robustness | The intrinsic compliance and flexible structure |
| | Fatigue avoidance and safety | No mechanical forces by robotic device on impaired hand. Light in weight |
| | Device adaptability to patient | Adaptable in terms of positioning and left/right hemiparetic upper limb with different sizes |
| | Control interface adaptability to patient | Auto-tuning calibration to better match the user-dependent nature of the EMG signal |
| | Device coupling with human arm | Devices firm grip at forearm by coupling of two rigid parts of support base having silicon skin to increase friction and two parallel velcro strap to tight both parts at the arm |
| | Object shape adaptability | Underactuation and passive compliance |
| | Mechanical power | Actuator with Controllable mechanical power (torque) |
| | Configurability | Modular design, easy modules assembly (flexible part slides in stiff part). Selectable desired control interface, i.e., EMG or push button |
| | Functional versatility | Devices functionality ranging from clinical needs to various indoor and outdoor ADL tasks |
| | Energy efficiency | Rechargeable light weight portable batteries. Actuator capability to maintain the stall torque at minimum current consumption |
| | Error tolerance | Underactuation, passive compliance and high friction through silicon skin at possible contact points |
| | Simple and intuitive interfaces | Trigger signal based simple control, LEDs mounted visual feedback control board for further intuitive interface |

## 4.6 Conclusion

This chapter presents design, analysis, manufacturing, experimental characterization and evaluation of two prototypes of soft robotic extra fingers that can be used as grasp compensatory robotic devices for hemiparetic upper limb. An approach is illustrated to to model the kinematics and the passive stiffness of flexible grippers and its realization in order to follow a desired trajectory for the finger. It explored the role of stiffness in grasping performance and demonstrated a novel method which is applicable to any tendon-driven, underactuated and passively compliant hand. The grasping performance of such grippers mainly depends on their intrinsic characteristics, e.g., passive joint compliance, instead of relying on active control for compliance used in complex manipulators. Firstly, a procedure to determine suitable joints stiffness is defined and then a possible realization in robotics fingers hardware structure is proposed. Without any loss of generality, It proposes a modular approach to define robotic hands. In most of the solutions existing in the literature, the hand is composed of a series of identical fingers, and the modularity is exploited at the finger level while we propose each phalanx built with same modules, i.e., a flexible joint and a rigid link. The model is built using theory of mechanics of tendon-driven hands coupled with beam theory. The feasibility of the approach was demonstrated through a framework composed of both simulations and exploitation of model in the realization of the Soft-SixthFinger. The experiments revealed that modelling joint stiffness and its realization using the proposed method improved the grasping performance and shape adaptation to a wider set of objects. Obviously, the presented approach can also reduce the manufacturing iterations needed to optimize the joints stiffness. Moreover, the proposed mathematical framework can also be used in continuum hands modelling (e.g., RBO hand) by further discretize the joints and links distribution. The devices are developed using rapid prototyping 3D printer and moulding techniques. the devices are tested with chronic stroke patients through qualitative experiments based on ADL.

# Chapter 5
# Wearable Sensory Motor Interfaces for Supernumerary Robotic Fingers

*Touch comes before sight, before speech. It is the first language and the last, and it always tells the truth*

*Margaret Atwood, Der blinde Mörder*

This chapter presents the design and development a new generation of extremely wearable haptic interfaces, acting as sensorimotor interfaces between the human wearer and the supernumerary robotic fingers. The small size and the specific shape of the developed devices guarantees an easy and intuitive wearability and leave the user's fingertips bare in order to not constraint the user's interaction with the surrounding environment in any way. Based on psychophysics evaluations, our novel interfaces deliver information about the cooperative task being performed (e.g., quality of the grasp) and the state of the extra-finger(s) where they are expected, providing the subject with a direct and co-located perception of the haptic feedback. Although the richness of the delivered information is very important, we strongly believe that sensorimotor consistency is crucial to design natural and embodied interfaces. In particular, two kinds of interfaces namely "vibrotactile ring" and hRing are proposed. The human user is able to control the motion of the robotic finger through switchs placed on rings, while being provided with vibrotactile and cutaneous feedback about the forces exerted by the robotic finger on the environment. To understand how to control the vibrotactile and cutaneous interface to evoke the most effective sensations, we executed perceptual experiments to evaluate its absolute and differential thresholds. We also carried out experiments with the subjects and haptic feedback significantly improved the performance in task execution in terms of completion time, exerted force, perceived effectiveness and wearability.



I. Hussain and D. Prattichizzo, *Augmenting Human Manipulation Abilities with Supernumerary Robotic Limbs*, Biosystems & Biorobotics 26,
https://doi.org/10.1007/978-3-030-52002-1_5

## 5.1 Introduction

Beyond design guidelines, another interesting issue is how to interface the extra finger with the human hand motion. In [23], we presented a possible control strategy able to transfer to one or more extra fingers a part or the whole motion of the human hand. We considered an extension of the mapping method proposed in [52, 55] to the case of a human hand augmented with a robotic extra finger. A commercial dataglove was used to measure the hand configuration during a grasping task. Although this control approach guarantees a reliable tracking of the human hand and can be extended to more fingers, there are two main drawbacks to be solved. First, the user lacks a feedback of the robotic finger status and can only perceive the force exerted by the device mediated by the grasped object. The second problem is related to the approaching phase of the grasp. In fact, the algorithm presented in [23] considers the motion of the whole hand to compute the motion of the extra finger, thus limiting the possibility of the user to make fine adjustments to adapt the finger shape to that of the grasped object.

## 5.2 Vibrotactile Haptic Feedback for Intuitive Control of Supernumerary Robotic Fingers

We firstly addressed these issues by introducing a vibrotactile interface that can be worn as a ring as shown in Fig. 5.1 [26, 36, 126, 127]. The human user receives information through the vibrotactile interface about the robotic finger status in terms of contact/no contact with the grasped object and in terms of force exerted by the device. Haptic stimuli have been indeed proved to enhance the performance of robotic systems in many scenarios [128–130]. However, most robotic systems with force reflection provide force feedback through grounded haptic devices, such as the Omega (Force Dimension, Switzerland) interfaces. Although these devices can be very accurate and able to provide a wide range of forces, their form factor makes them not suitable to be used in wearable applications, where the system needs to be lightweight, small, and easy to wear [131]. For this reason, we designed a custom wearable vibrotactile interface, since vibrotactile stimuli convey rich information and have an extremely compact form factor with respect to more popular grounded interfaces [132–134].

Regarding the grasp approaching phase, we introduce a new control strategy that enables the finger to autonomously adapt to the shape of the grasped object. We developed a new finger prototype with 3 DoFs and three modules to resemble the kinematics of the human finger. Each module has been equipped with a force sensor able to detect contacts with the grasped object. We defined a grasping procedure that starts from a predefined position, when the robotic finger is open at its maximum. As soon as one module is in contact with the object, that module stops its motion, while the others keep moving toward the object. The details of the closing procedure will

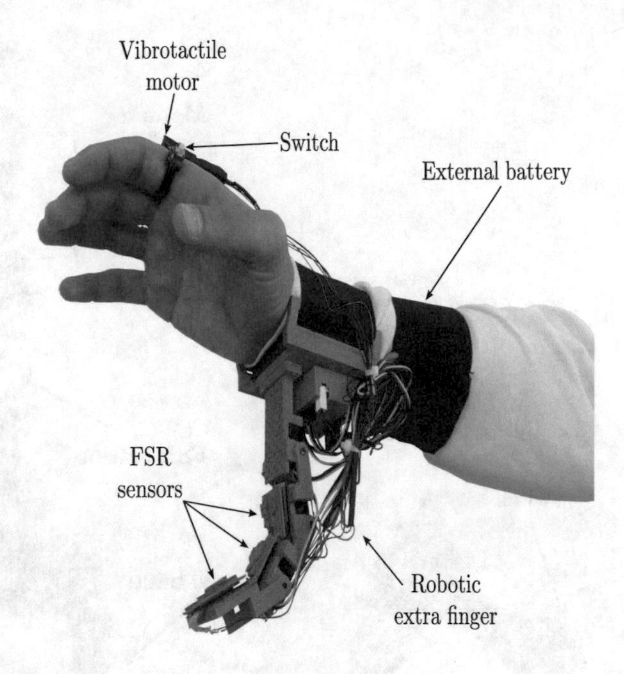


**Fig. 5.1** The robotic extra finger together with the vibrotactile interface ring. The ring provides haptic feedback through a vibrating motor and enables the user to start and stop the finger motion through a switch

be described in Sect. 5.2.1. The procedure can be activated by pressing the switch placed on the interface ring (see Fig. 5.1). This approach dramatically simplifies the interaction with the robotic finger, reducing it to the activation of a grasping procedure through the wearable switch. This shared autonomy between the user and the device has been kept also in the control of the exerted force. We developed a variable compliance control that let the finger adapt its stiffness according to the force necessary to guarantee a stable grasp. The user is provided through the vibrating ring with information about the exerted forces and, therefore, about the finger compliance.

We tested the system in a pick and place task where users were asked to move an object between two predefined positions by taking advantage of the extra finger. We demonstrated that haptic feedback, together with the intuitive human robot interface, enhances users' performance in terms of completion time and exerted force. The rest of the paper is organized as it follows. Section 5.2.1 describes the extra finger and the ring interface. Section 5.2.3 presents two preliminary experiments aiming at evaluating the absolute and differential thresholds of our vibrotactile haptic device. Section 5.2.4 deals with the pick-and-place experiment carried out to evaluate the effectiveness of the system, while in Sect. 5.2.5 conclusion and future work are outlined.

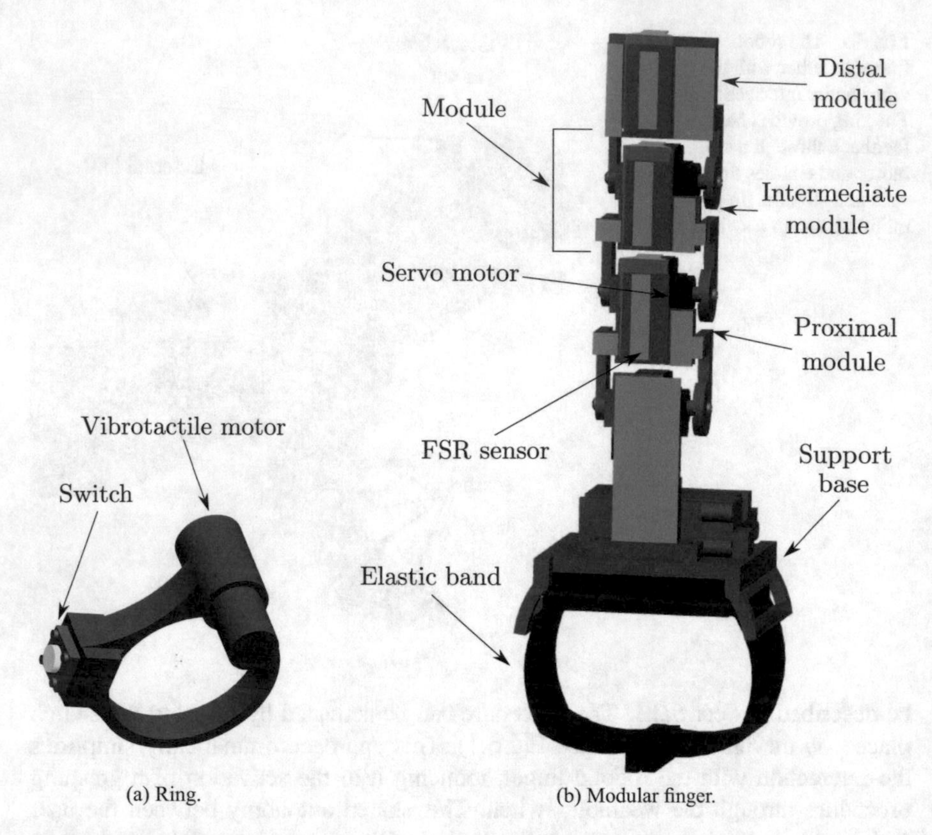


(a) Ring. (b) Modular finger.

**Fig. 5.2** CAD models of the vibrotactile interface and of the modular finger. Three single DoF modules are connected to a wrist elastic band. The ring is equipped with a push button which is used as an interface with the user and a vibrotactile motor to provide an haptic stimulus

## *5.2.1 Sensorized Supernumerary Robotic Finger*

The robotic finger presented in Chap. 2 is modified to equip the modules with force sensors and position feedback.

In particular, a Force Sensing Resistor (FSR) (408, Interlink Electronics Inc., USA) is placed on each module, as reported in Fig. 5.2. These sensors will be used for the closing procedure and the compliance variation described in Sect. 5.2.2. The servo motor of each module is modified in order to access the reading of internal potentiometer for the actual position of the motor.

The vibrotactile ring interface is designed to be worn on the index finger of the human hand, see Fig. 5.1. The ring is equipped with a switch and a vibro motor, as shown in Fig. 5.2. The ring housing is 3D printed. The motor used is an eccentric rotating mass vibrotactile motor (Precision MicroDrives, United Kingdom). The switch is used to start the finger closing procedure, as described in Sect. 5.2.2, and move the finger back to the initial grasp position. The vibrotactile motor is used to provide vibrotactile feedback, as explained in Sect. 5.2.4.

### 5.2.2 Grasping Procedure and Compliance Regulation

The user can command the finger motion by using the wearable switch placed on the ring. When the switch is activated, the finger starts to close with a fixed joint angle increment, equal in each module, from a predefined position. We considered the completely extended finger as the starting position to enlarge the set of possible graspable objects.

During the grasping phase, the FSR sensors are in charge of detecting the contact of each module with the grasped object. In order to have suitable contact points, we set different closing priorities for each module. If the distal module (see Fig. 5.2b) comes in contact first, the remaining two modules stop. If the intermediate module gets in contact first, only the proximal one stops, while the distal module keeps moving. Finally, if the proximal module comes in contact first, two different behaviors can occur: (1) the intermediate module gets in contact first and the distal one is free to move to get in contact with the object, (2) the distal module comes in contact first and then also the intermediate one stops. The grasping procedure is commanded by the user acting on the switch. When the grasp is complete, the finger starts to autonomously keep the grasp stable. This grasp stabilization is obtained by controlling the compliance of each module. The compliant behavior of a module with different values of the scaling factor $k_d$ is reported in Fig. 5.3. The force at each module is measured by using the FSR sensor. The compliance decreases with the increase of $k_d$. The basic idea is that the module can change its compliance according to the force observed by each module through the relative FSR sensor. Thus, when the user pushes the object toward the extra finger to tight the grasp, the device becomes stiffer. The possibility to independently regulate each module's compliance allows to adapt the finger to the the shape of the grasped object also during manipulation tasks. Similar to what we did for the grasping procedure, we set the same priorities between the three modules also regarding the compliance variation. When the user wants to release the grasped object, he just needs to lower the force exerted by his hand on the object and, automatically, the robotic device will make its joints more compliant. Eventually, by pressing again the switch, the robotic finger moves back to its home position by following a predefined trajectory.

### 5.2.3 Absolute and Differential Thresholds

In order to understand how to drive the vibrotactile ring correctly to evoke the most effective cutaneous sensations, we ran two preliminary experiments aiming at evaluating the absolute and differential thresholds of our device. The motor we use is an

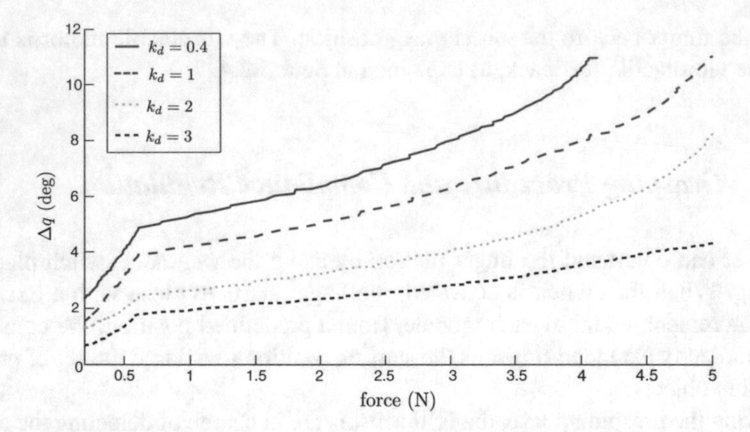


**Fig. 5.3** Compliance behavior of a single module with different value of $k_d$. Force vs displacement is plotted. The force is measured using FSR sensor and the displacement is recorded by the position feedback of the modified servo motor

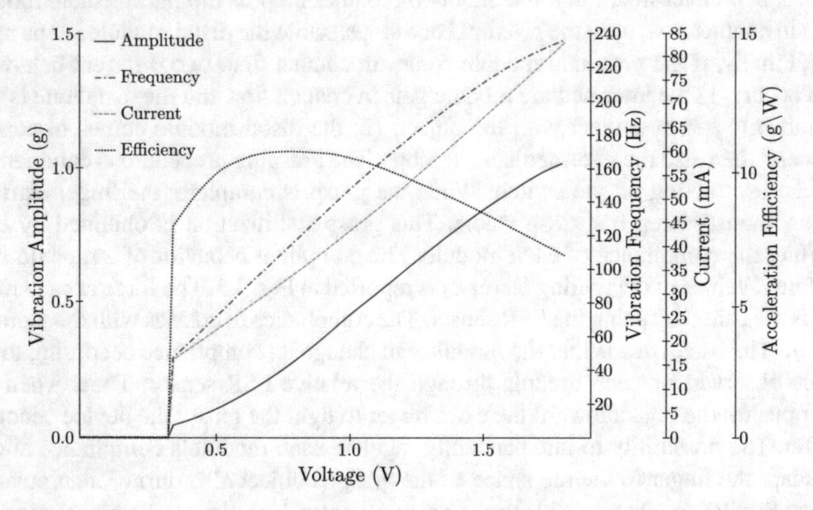


**Fig. 5.4** Performance characteristics of the precision microdrives 4 mm vibration motor (11 mm type)

eccentric rotating mass vibrotactile motor, and it is not possible to separately control amplitude and frequency of vibration. The relationship between input voltage, and the amplitude and frequency of the vibration is shown in Fig. 5.4. Subjects were required to wear the vibrotactile ring on their right index proximal phalanx. Moreover, to avoid any additional cue, subjects were blindfolded and wore noise-canceling headphones.

The absolute threshold can be defined as the "smallest amount of stimulus energy necessary to produce a sensation" [135], and it gives us information about the smallest vibration we need to provide in order to produce a perceivable sensation by the human

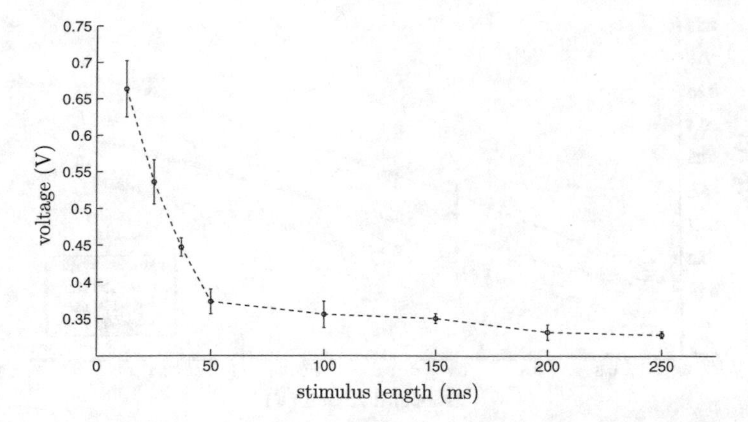


**Fig. 5.5** Absolute threshold. Mean values and standard errors of the mean (SEM) are plotted. The relationship between input voltage, and the amplitude and frequency of the vibration can be found in Fig. 5.4

user. Six participants took part in the experiment. The experimenter explained the procedures and spent about three minutes adjusting the setup to be comfortable before the subject began the experiment.We evaluated the absolute threshold by using the simple up-down method [136]. We used a step-size of 0.08 V, which reduced of 0.02 V every reversal. We considered the task completed when four reversals occurred.

Subjects were required to wear the cutaneous device as shown in Fig. 5.1 and tell the experimenter when they felt the stimulus. Each participant performed forty-eight repetitions of the simple up-down procedure, with six repetitions for each considered duration of the vibratory stimulus: 13, 25, 37, 50, 100, 150, 200, and 250 ms. Figure 5.5 shows the absolute thresholds averaged over all subjects.

The differential threshold can be in turn defined as "the smallest amount of stimulus change necessary to achieve some criterion level of performance in a discrimination task" [135]. This gives us information about how much different two vibrations provided with our device need to be in order to be perceived as different by the human user. This threshold is often referred to as just-noticeable difference or JND. The differential threshold of a perceptual stimulus reflects also the fact that people are usually more sensitive to changes in weak stimuli than they are to similar changes in stronger or more intense stimuli. The German physician Ernst Heinrich Weber suggested the simple proportional law $JND = kI$, suggesting that the differential threshold increases by increasing the intensity $I$ of the stimulus. Constant $k$ thus referred to as "Weber's fraction". The experimental setup was the same as described in Sect. 5.2.3. Nine participants took part in this experiment. As in Sect. 5.2.3, the experimenter explained the procedures and spent about three minutes adjusting the setup to be comfortable before the subject began the experiment. We evaluated the differential threshold using again the simple up-down method [136]. We used again a step-size of 0.08 V, which reduced of 0.02 V at every reversal. We considered the task completed when four reversals occurred. Subjects were required to wear the

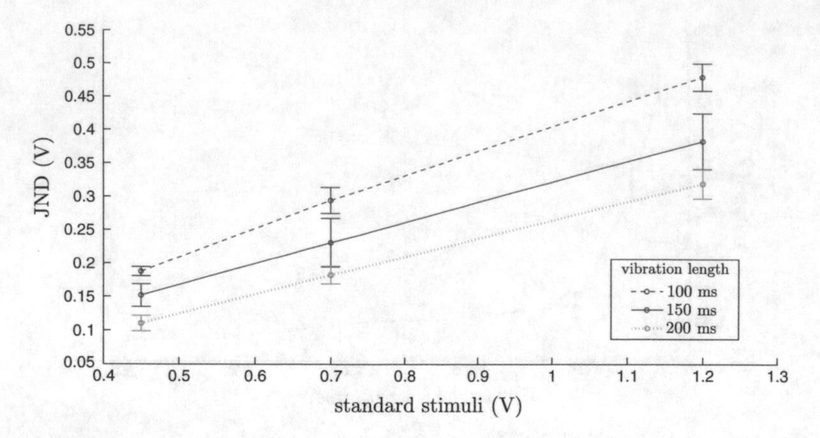


**Fig. 5.6** Differential threshold. Mean values and standard errors of the mean (SEM) are plotted. The relationship between input voltage, and the amplitude and frequency of the vibration can be found in Fig. 5.4

cutaneous device and tell the experimenter when the two vibrations provided felt different. We tested the JND at three reference stimuli: 0.45, 0.7, and 1.2 V, which corresponded, respectively, to vibrations of amplitude 0.15, 0.30, and 0.62 g, and of frequency 68, 100, and 160 Hz (see Fig. 5.4). We considered also three different vibration lengths: 100, 150, and 200 ms. We did not consider lengths < 100 ms to be sure that everyone would be able to perceive them at all reference stimuli (see Fig. 5.5). The results observed are in agreement with previous results in the literature [137]. Each participant performed eighteen trials of the up-down procedure, with two repetitions for each of reference stimulus and vibration length considered. Figure 5.6 shows the differential thresholds averaged over all subjects. For the reference stimuli of 0.45, 0.7, and 1.2 V, and a vibration length of 100 ms, the average JNDs are 0.19, 0.29, and 0.48 V, respectively. Thus, the Weber fractions are 0.42, 0.42 and 0.40, respectively, following Weber's Law. For the reference stimuli of 0.45, 0.7, and 1.2 V, and a vibration length of 150 ms, the average JNDs are 0.15, 0.23, and 0.38 V, respectively. Thus, the Weber fractions are 0.34, 0.33 and 0.32, respectively, following Weber's Law. Finally, for the reference stimuli of 0.45, 0.7, and 1.2 V, and a vibration length of 200 ms, the average JNDs are 0.11, 0.18, and 0.32 V, respectively. Thus, the Weber fractions are 0.25, 0.26 and 0.27, respectively, following Weber's Law.

### 5.2.4 Experimental Evaluation

In order to evaluate the effectiveness of our extra finger device and the usefulness of vibrotactile haptic feedback, we carried out a pick-and-place experiment. The experimental setup was composed of our robotic extra finger device, the vibrotactile

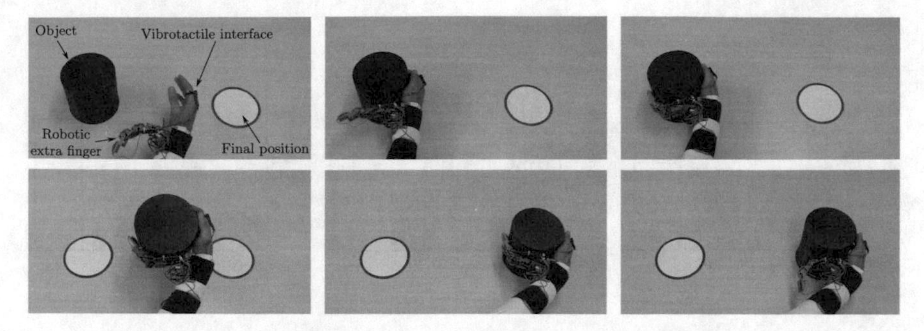


**Fig. 5.7** Experimental setup. The task consisted of grasping the cylinder from its starting position, lifting it from the table, and moving it to its final position, being as fast as possible. The cylinder had a radius of 7.5 cm and a height of 15 cm

ring, and a cylinder, as shown in Fig. 5.7. The cylinder was made of ABSPlus, it had a radius of 7.5 cm, a height of 15 cm, and a weight of 150 g. The robotic extra finger device was controlled as detailed in Sect. 5.2.2. The task consisted of grasping the cylinder, lifting it from the table, and moving it 45 cm right, being as fast as possible. The initial and final positions of the cylinder were marked on the table by two red circles (see Fig. 5.7). Ten participants took part in the experiment. The experimenter explained the procedures and spent about five minutes adjusting the setup to be comfortable before the subject began the experiment. Each participant made twenty randomized trials of the pick-and-place task, with five repetitions for each feedback condition proposed:

- vibration bursts on making and breaking contact with the grasped object and when close to the actuators' force limit (condition $V_{mb}$),
- vibration bursts on the intensity of the force exerted by the robotic finger (condition $V_t$),
- continuous vibrations proportional to the intensity of the force exerted by the robotic finger (condition $V_c$),
- no haptic feedback at all (condition N).

In condition $V_{mb}$, the vibrotactile ring provided a 200 ms-long vibration on making and breaking contact with the cylinder. Moreover, it also provided a 200 ms-long vibration when the force exerted by the robotic finger was close to the maximum force that actuators could provide at the fingertip (i.e., 5 N). The amplitude and frequency of these vibrations were set to 0.30 g and 100 Hz, respectively, so to be easy to recognize (input voltage $v_i = 0.7$ V, see Sect. 5.2.3). In condition $V_t$, the vibrotactile ring provided 200 ms-long vibrations on making and breaking contact with the cylinder and when the force sensed by the robotic finger was equal to 2 N and 4 N (we considered the maximum force sensed among the three sensors). On the making and breaking contact, the amplitude and frequency of the vibrations were set to 0.30 g and 100 Hz, respectively, as in condition $V_{mb}$ (input voltage $v_i = 0.7$ V). On the other hand, when the force sensed by the robotic finger was 2 N and 3 N,

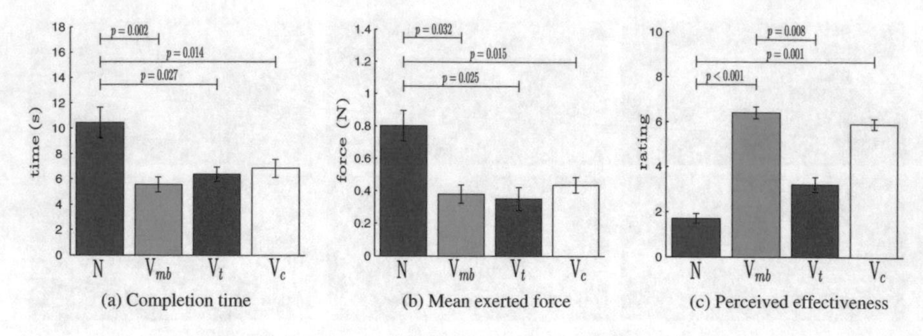


(a) Completion time (b) Mean exerted force (c) Perceived effectiveness

**Fig. 5.8** Pick-and-place experiment. Completion time, mean exerted force, and perceived effectiveness for conditions providing vibration bursts on making and breaking contact and when close to the actuators' force limit ($V_{mb}$), vibration bursts on the intensity of the force exerted ($V_t$), continuous vibrations proportional to the intensity of the force exerted ($V_c$), and no haptic feedback (N). Mean and Standard Error of the Mean (SEM) are plotted

the vibrotactile motor was provided with inputs $v_i$ of 0.9 V (amp. 0.43 g, freq. 124 Hz) and 1.7 V (amp. 0.94 g, freq. 222 Hz), respectively. Amplitude values and force thresholds were chosen to be easy to distinguish (see Sect. 5.2.3) and fit the vibrotactile motor specifications (see Fig. 5.4).

In condition $V_c$, the vibrotactile ring provided continuous vibrotactile feedback proportional to the intensity of the force exerted by the robotic finger. The commanded input voltage $v_i$, proportional to the mean force sensed on the robotic finger, was evaluated as

$$v_i = \frac{(f_{e,max} + 0.3)}{2.5},$$

where $f_{e,max}$ is the maximum force registered among the three sensors on the robotic finger. The relationship between input voltage $v_i$ and the amplitude and frequency of the vibration can be found in Fig. 5.4. For example, when the sensors register an average force of 2 N, the vibrotactile ring provides a vibration of amplitude 0.43 g and frequency 124 Hz (as in condition $V_t$). No vibration was provided when the robotic finger was not in contact with the object. In condition N, no haptic feedback was provided. Completion time and mean exerted forces provided a measure of performance. A low value of these metrics denotes the best performance. The time started to be recorded when the semi-autonomous grasping phase was completed and the user lifted the object, and it stops when the object was placed on its final position. Fig. 5.8a shows the average task completion time. All the data passed the Shapiro-Wilk normality test and the Mauchly's Test of Sphericity. A repeated-measure ANOVA showed a statistically significant difference between the means of the four feedback conditions ($F(3,27) = 16.597$, $p < 0.001$, $a = 0.05$). Post hoc analysis with Bonferroni adjustments revealed a statistically significant difference between condition N and all the others (N vs. $V_{mb}$, $p = 0.002$; N vs. $V_t$, $p = 0.027$; N vs. $V_c$, $p = 0.014$).

Figure 5.8b shows the average exerted force, evaluated as the mean over time of $f_{e,max}$. All the data passed the Shapiro-Wilk normality test. Mauchly's Test of Sphericity indicated that the assumption of sphericity had been violated ($\chi^2(2) = 17.008$, $p < 0.001$). A repeated-measure ANOVA with Greenhouse-Geisser correction showed a statistically significant difference between the means of the four feedback conditions (F(1.482,13.340) = 12.015, $p = 0.002$, a = 0.05). Post hoc analysis with Bonferroni adjustments revealed again a statistically significant difference between condition N and all the others (N vs. $V_{mb}$, $p = 0.032$; N vs. $V_t$, $p = 0.025$; N vs. $V_c$, $p = 0.015$). In addition to the quantitative evaluation reported above, we also measured users' experience. Immediately after the experiment, subjects were asked to report the effectiveness of each feedback condition in completing the given task using bipolar Likert-type seven-point scales. Fig. 5.8c shows the perceived effectiveness of the four feedback conditions. A Friedman test showed a statistically significant difference between the means of the four feedback conditions ($\chi^2(3) = 27.903$, $p < 0.001$). The Friedman test is the non-parametric equivalent of the more popular repeated measures ANOVA. The latter is not appropriate here since the dependent variable was measured at the ordinal level. Post hoc analysis with Bonferroni adjustments revealed a statistically significant difference between conditions N and $V_{mb}$ ($p < 0.001$), N and $V_c$ ($p = 0.001$), and $V_t$ and $V_{mb}$ ($p = 0.008$). Moreover, although condition $V_t$ was not found significantly different from condition $V_c$, comparison between them was very close to significance ($p = 0.072$). Finally, subjects were asked to choose the condition they preferred the most. Condition $V_{mb}$ was preferred by six subjects and condition $V_c$ was preferred by four subjects.

### 5.2.5 Conclusion

The first part of this chapter presented the integration of the wearable robotic extra finger with a ring able to provide vibrotactile haptic feedback. We introduced also a new solution for the grasping phase based on a wearable switch embedded in the ring and a closing policy that let the robotic finger adapt to the object's shape. In order to evaluate the effectiveness of our extra finger device and the usefulness of vibrotactile haptic feedback, we carried out a pick-and-place experiment. Ten subjects were asked to grasp a cylinder, lift it, and move it 45 cm right, being as fast as possible. Haptic feedback significantly improved the performance of the task in terms of completion time, exerted force, and perceived effectiveness. Moreover, all subjects preferred conditions employing haptic feedback. Within the conditions employing haptic feedback, no statistical difference was found in terms of completion time and force exerted. However, conditions $V_{mb}$ and $V_c$ were perceived as more effective than $V_t$. This may be due either to a saturation effect on the skin receptors or to the fact that the very rich information provided in condition $V_t$ can be difficult to be understood by the user.

## 5.3 The hRing as a Wearable Haptic Interface for Supernumerary Robotic Fingers

The wearable electronics business has powered over $14 billion in 2014 and it is estimated to power over $70 billion by 2024. However, the proposed vibrotactile interface and commercially-available wearable devices still provide very limited haptic feedback, mainly focusing on vibrotactile sensations. Towards a more realistic feeling, the second part of the chapter proposes a novel wearable cutaneous device for the proximal finger phalanx, called "hRing". It consists of a vibro motor and two servo motors that move a belt placed in contact with the user's finger skin. When the motors spin in opposite directions, the belt presses into the user's finger, while when the motors spin in the same direction, the belt applies a shear force to the skin.

Researchers have recently proposed several wearable haptic devices able to provide rich cutaneous sensations at the fingertips [131, 138], enabling a compelling interaction with virtual and remote objects. However, although quite effective, placing such haptic-enabled wearable devices at the fingertips may significantly affect the quality of the fingertip tracking, especially if done using unobtrusive solutions such as the Microsoft Kinect Sensor or the Leap Motion Controller. Moreover, wearable cutaneous devices prevents the users from using their fingertips to interact with real objects(e.g., in the case of augmented reality scenarios). Towards a realistic and unobtrusive feeling of interacting with real and virtual objects, we propose a novel wearable cutaneous device for the proximal finger phalanx, called "hRing", shown in Fig. 5.9a.

A preliminary version of this wearable device has been presented in [126]. With respect to the version proposed in [126], the new hRing presented in this work offers enhanced mobility of the finger phalanx, ambidexterity, a vibrotactile motor, and a lighter body structure. Pacchierotti et al. [126] used the earlier version of this interface to interact with objects in a virtual environment and results show that providing cutaneous feedback through the proposed device improved the performance and perceived effectiveness of the considered virtual pick-and-place task of 20 and 47% with respect to not providing any force feedback, respectively. In this work, we propose the hRing as interface to interact with a novel underactuated soft robotic finger (SRF), as shown in Fig. 5.9.

### 5.3.1 *The hRing Structure*

Wearability is the key concept in the design of the proposed hRing cutaneous device. Besides wearability, other essential features are comfort, effectiveness, and ease of use. For all the reasons mentioned above, the hRing device has been designed to be worn on the proximal part of the index finger.

Figure 5.10 shows all the parts of hRing device. A strap band is used to secure the device on the finger. The static platform and the pulleys are realized in Acrylonitrile

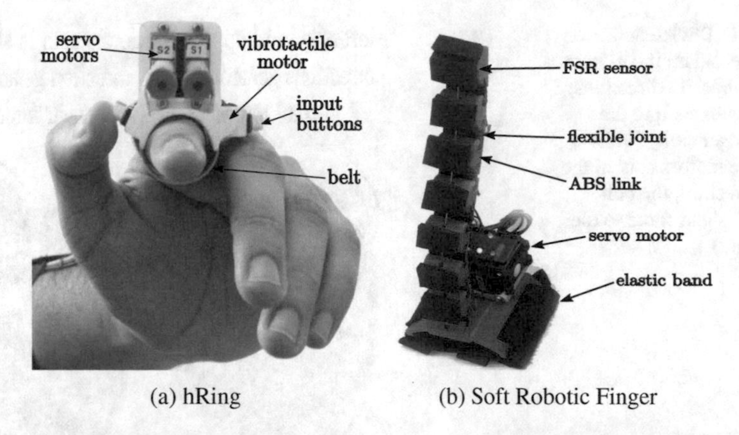


(a) hRing (b) Soft Robotic Finger

**Fig. 5.9** The hRing and the soft robotic finger (SRF)

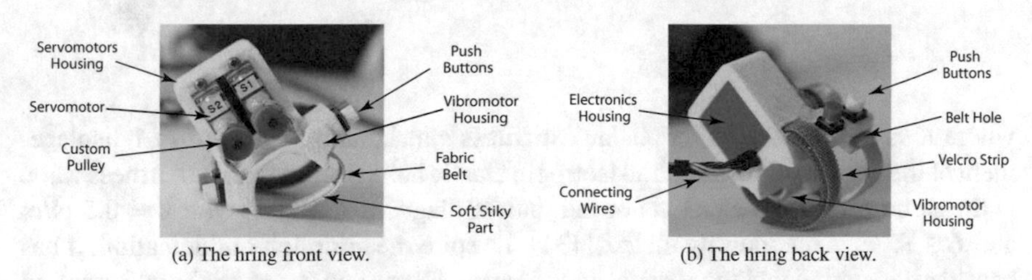


(a) The hring front view. (b) The hring back view.

**Fig. 5.10** The complete parts of the hring device

Butadiene Styrene (ABS-Plus, Stratasys, USA) through the use of a commercial 3D printer. The servomotors are HS-40 Microservo (HiTech, Republic of Korea). Each can provide a maximum torque of 0.05 Nm.

The working principle of the device is depicted in Fig. 5.11. Similarly to the principle proposed by Minamizawa et al. [138], when the two motors rotate in opposite directions, the belt is pulled up, providing a force normal to the finger (left side of Fig. 5.11). On the other hand, when motors spin in the same direction, the belt applies a shear force to the finger (right side of Fig. 5.11).

The servomotors are position controlled, which means that it is only possible to command a desired angle. The relationship between the commanded angle $\theta$ expressed in radians and the belt displacement $d$, due to a single motor, is

$$d = r\theta, \tag{5.1}$$

where $r = 5$ mm is the radius of the servo motor pulley. To relate the belt displacement $d$ to the force applied on the finger proximal phalanx $\boldsymbol{f} \in \mathbb{R}^3$, we assume

$$\boldsymbol{f} = K\boldsymbol{p}, \tag{5.2}$$

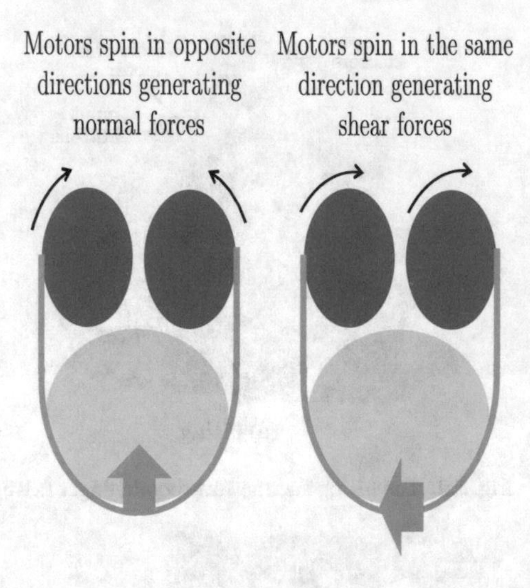


**Fig. 5.11** Device actuation principle. When the motors spin in opposite directions, the belt presses into the user's finger (left), while when the motors spin in the same direction, the belt applies a shear force to the skin (right)

where $K \in \mathbb{R}^{3\times 3}$ is the finger phalanx stiffness matrix and $p$ is the normal displacement of the belt. We considered an isotropic elastic behavior, so that the stiffness value is the same for all the elements of the matrix diagonal $K = kI$, with $k = 0.5$ N/m and $I \in \mathbb{R}^{3\times 3}$ is the identity matrix [131]. Despite the simplicity of actuation, it has been demonstrated that the vertical and shearing forces generated by the deformation of the fingerpads can reproduce reliable weight sensations even when proprioceptive sensations on the wrist and the arm are absent [139]. Our objective is to understand if these type of stimuli are still effective when provided to the proximal phalanx of the finger instead of the fingertip.

## 5.3.2 Differential Thresholds

To understand how to correctly modulate the cutaneous stimuli provided, we carried out two preliminary experiments evaluating the differential thresholds for normal and shear stimuli. The differential threshold can be defined as "the smallest amount of stimulus change necessary to achieve some criterion level of performance in a discrimination task" [135]. It gives us information about how different two displacements provided with our device need to be in order to be perceived as different by a human user. This threshold is often referred to as just-noticeable difference or JND. The differential threshold of a perceptual stimulus reflects also the fact that people are usually more sensitive to changes in weak stimuli than they are to similar changes in stronger or more intense stimuli. The German physician Ernst Heinrich Weber proposed the simple proportional law $JND = kI$, suggesting that the differential threshold increases with increasing the stimulus intensity $I$. Constant $k$ is thus referred to as "Weber's fraction". For example, Schorr et al. [140] measured the ability of users to discriminate environment stiffness using varying levels of skin stretch at the finger pad. Results showed a mean Weber fractions of 0.168. Similarly, Guinan et al. [141] found a mean Weber fraction of 0.2 for their skin stretch sliding plate tactile device.

Seven participants took part in the experiments, including one woman and six men. Four of them had previous experience with haptic interfaces. None of the participants reported any deficiencies in their visual or haptic perception abilities, and all of them were right-hand dominant. Subjects were required to wear the device on their right index proximal phalanx. To avoid providing any additional cue, subjects were blindfolded and wore noise-canceling headphones.

### 5.3.3 *Normal Stimuli*

We evaluated the differential threshold for normal stimuli using the simple up-down method [136]. We used a step-size for the servo motors of $\alpha = 1°$, that corresponded to a normal displacement of the belt of $p = 0.09$ mm (see Eq. (5.1)). We considered the task completed when six reversals occurred. Subjects were required to wear the cutaneous device and tell the experimenter when the two stimuli provided felt different. We tested the JND at three standard stimuli: 1, 2, 3, and 4 mm of displacement into the finger pad. Each participant performed eight trials of the simple up-down procedure, with two repetitions for each standard stimulus considered. Fig. 5.12a shows the differential thresholds registered for each reference stimulus. For the reference stimuli of 1, 2, 3, and 4 mm, the average JNDs are 0.07, 0.12, 0.16, and 0.19 mm. respectively. Thus, the Weber fractions are 0.07, 0.06, 0.05, 0.05, respectively, following Weber's Law.

We evaluated the differential threshold for shear stimuli using the simple up-down method again [136]. We used a step-size for the servo motors of $\alpha = 1°$, that corresponds to a lateral movement of the belt on the finger pad of 0.09 mm (see Eq. (5.1)). We considered the task completed when six reversals occurred. Subjects were required to wear the cutaneous device and tell the experimenter when the two stretches provided felt different. We tested the JND at three standard stimuli: 0.45, 0.90, 1.35, and 1.80 mm of stretch on the finger pad. The normal displacement into the skin was fixed to 6 mm. Each participant performed eight trials of the simple up-down procedure, with two repetitions for each standard stimulus considered. Figure 5.12b shows the differential thresholds registered for each reference stimulus. For the reference stimuli of 0.45, 0.90, 1.35, and 1.8 mm, the average JNDs are 0.08, 0.15, 0.17, and 0.26 mm, respectively. Thus, the Weber fractions are 0.18, 0.16, 0.13, 0.15, respectively, following Weber's Law. No slippage between the belt and the skin took place during the trials.

We integrated the "hring" with the soft robotic finger shown in Fig. 5.9. The detailed design of the robotic finger is presented in Sect. 4.

Two chronic stroke patients (male, age 35 and 56) took part to our experimental evaluation on how the proposed integrated robotic system can be used for hand

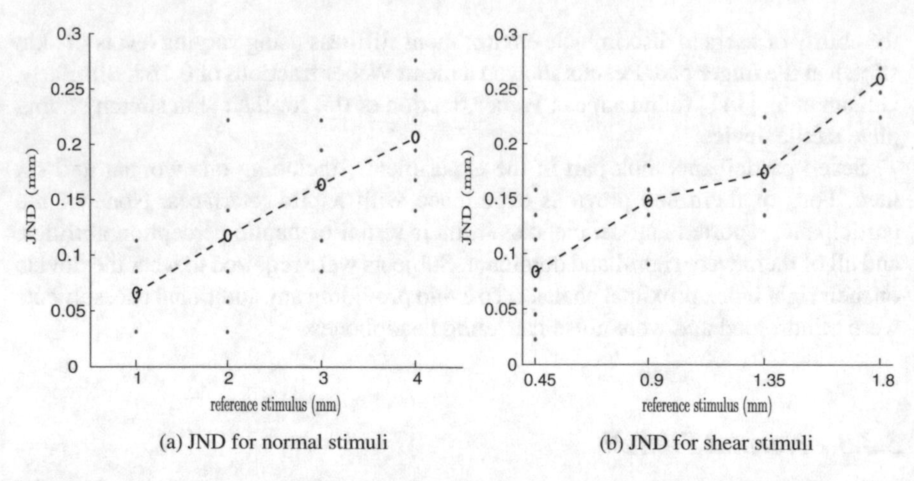


(a) JND for normal stimuli (b) JND for shear stimuli

**Fig. 5.12** Differential threshold. Mean values are plotted as empty black circles, thresholds for each subject are plotted as blue dots

**Fig. 5.13** The integrated system used by a patient in Activities of Daily Living (ADL)

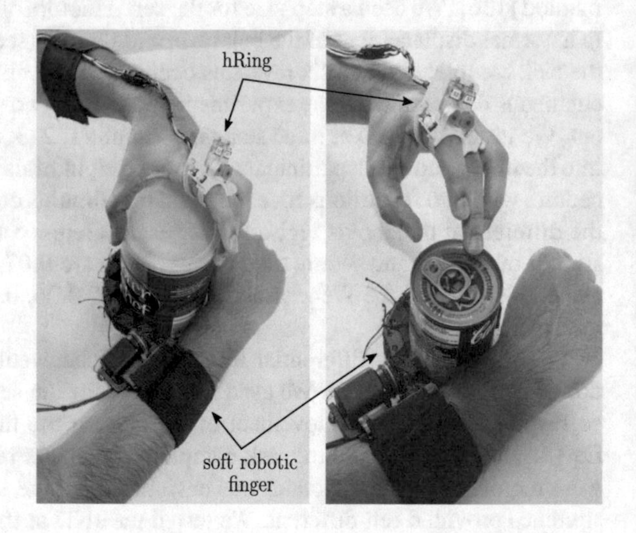


grasping compensation. The subjects were required to wear the robotic finger on their paretic hand and the hRing on the healthy hand index finger (see Fig. 5.13). The proposed compensatory tool can be used by subjects showing a residual mobility of the arm. For being included in this experimental evaluation, patients had to score ≤2 when their motor functions were tested with the National Institute of Health Stroke Scale (NIHSS), item 5, "paretic arm". Moreover, the patients had to show the following characteristics: normal consciousness (NIHSS, item 1a, 1b, 1c = 0), absence of conjugate eyes deviation (NIHSS, item 2 = 0), absence of complete hemianopia (NIHSS, item 3 ≤ 1), absence of ataxia (NIHSS, item 7 = 0), absence of completely sensory loss (NIHSS, item 8 ≤ 1), absence of aphasia (NIHSS, item 9 = 0), absence of profound extinction and inattention (NIHSS, item 11 ≤ 1).

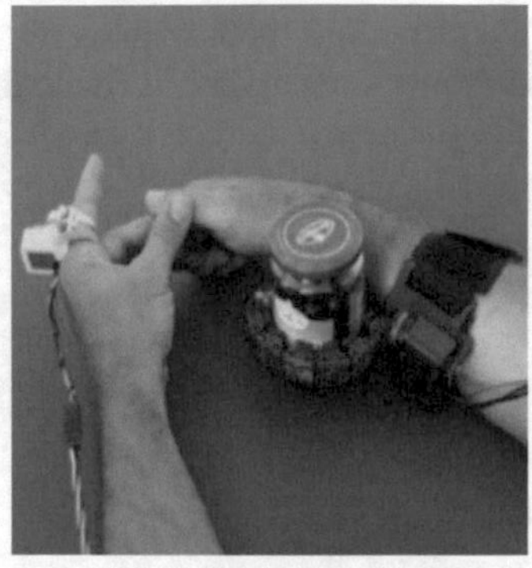

(a) Opening the jar of jam.

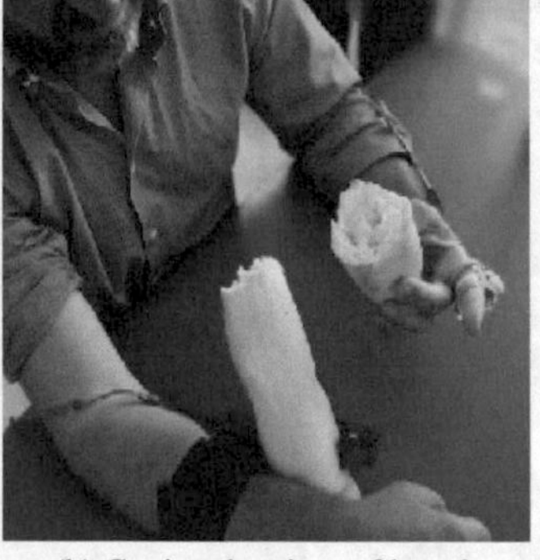

(b) Cutting the piece of bread.

**Fig. 5.14** Examples of other bimanual tasks

Each patient uses the soft finger for bimanual tasks typical in Activities of Daily Living (ADL), such as unscrewing a cap of tomato jar, opening a popcorn bag, or opening a can of beans. The hRing interface is used to both control the flexion/extension motion of the robotic finger and to provide the patient with haptic feedback about the forces exerted by the robotic finger on the environment (Fig. 5.14).

Pressing the blue button on the external side of the hRing (see Fig. 5.9) initiates the flexing procedure of the robotic finger. The finger will close until a contact with an object through the FSR mounted on its fingertip is detected. As soon as the contact happens, the robotic finger stops its flexion and the hRing generates a short vibration burst to notify the patient. If the patient presses the blue button again, the finger increases the grasping force on the object. During this process, the hRing belt squeezes the patient's finger proportionally to the grasp force applied by the robotic finger on the environment. If the grasping force exerted by the robotic finger reaches the maximum force applicable by the motor, a second vibration burst alerts the user. When the patient is satisfied with the grasping configuration, he or she can proceed with the task. Finally, pressing the yellow button on the hRing will initiate the opening procedure of the robotic finger. The hRing releases its belt and provides a vibration burst when the opening procedure is completed.

Both patients found the soft robotic finger useful in the considered ADL tasks. Moreover, they found the hRing intuitive to use and unobtrusive. Finally, they reported haptic feedback to be a valuable information to estimate the quality of the grasp. Both patients would like to be able to use it at home.

## *5.3.4 Conclusions*

In second part of this chapter, we briefly introduced the integration of the hRing as interface for a wearable robotic soft finger. The hRing enables patients to easily control the motion of the robotic finger while providing them with haptic feedback about the status of the grasping action. Two chronic stroke patients found the system useful for ADL tasks, the hRing easy to use, and the haptic feedback very informative.

# Chapter 6
# Wearable EMG Interfaces for Motion Control of Supernumerary Robotic Fingers

This chapter presents two kinds of electromyographic (EMG) control interfaces for supernumerary robotic fingers. In particular, frontalis muscle cap and arm EMG interface. The former is more suitable for underactuated soft fingers while the latter is developed to control the motion and compliance of fully actuated robotic finger.

In frontalis muscle cap the electrodes and acquisition boards are embedded in a cap which allows the user to control the device motion through wireless communication by contracting the frontalis muscle. We used the cap interface with soft sixth finger to perform qualitative experiments involving six chronic stroke patients. Results show that the proposed system significantly improves the performance of the considered tests and the autonomy in ADL.

The arm EMG interface composed of a commercial EMG armband for gesture recognition to be associated with the motion control and to regulate the compliance of the robotic device. On fully actuated robotic finger side, we proposed a control approach to regulate the compliance of the device through servo actuators and presented the updated version of the device where the adduction/abduction motion is realized through ball bearing and spur gears mechanism. We have validated the proposed interface with two sets of experiments related to compensation and augmentation. In the first set of experiments, different bi-manual tasks have been performed with the help of the robotic device and simulating a paretic hand since this novel wearable system can be used to compensate the missing grasping abilities in chronic stroke patients. In the second set, the robotic extra finger is used to enlarge the workspace and manipulation capability of healthy hands. In both sets, the same EMG control interface has been used. The obtained results demonstrate that the proposed control interface is intuitive and can successfully be used, not only to control the motion of a supernumerary robotic finger, but also to regulate its compliance. The proposed approach can be exploited also for the control of different wearable devices that has to actively cooperate with the human limbs.



I. Hussain and D. Prattichizzo, *Augmenting Human Manipulation Abilities with Supernumerary Robotic Limbs*, Biosystems & Biorobotics 26,
https://doi.org/10.1007/978-3-030-52002-1_6

## 6.1 Introduction

One of the major challenge in augmenting/compensating human capabilities through robotic extra limbs concerns the development of a suitable control interfaces for the integration of the device motion with that of the human. Wu and Asada [22] presented a control algorithm enabling a human hand augmented with two robotic fingers to share the task load together and adapt to diverse task conditions. Postural synergies were found for the seven-fingered hand comprised of two robotic fingers and five human fingers through the analysis of measured data from grasping experiments. In [23], a mapping algorithm able to transfer to an arbitrary number of robotic extra-fingers the motion of the human hand has been presented. The mapping algorithm was based on the definition of a virtual object obtained as a function of a set of reference points placed on the augmented hand (human hand and robotic fingers). The mapping algorithm allowed to move the extra-fingers according to the human hand motions without requiring explicit command by the user.

Both the approaches used an instrumented glove to track the human hand presenting some limitations which affected their practical application. Patients with a paretic hand cannot properly control finger motions, thus a dataglove interface cannot be used. The estimation of the human hand posture and fingers motion implies a reliable and computationally expensive hand tracking. Moreover, datagloves can be only used for position control of the robotic device without having any control on force or stiffness regulation. As a preliminary solution to the above mentioned issues, we implemented a trigger based control approach [26, 36]. The trigger signal was activated by a wearable switch placed on a ring. A single switch activation regulated the stop/motion of the finger along a predefined flexion trajectory, while a double activation switched from flexion to extension and viceversa. Although the ring based control approach resulted simple and intuitive, this control interface involved human hand thumb—on the same hand for healthy subjects [26] and on the controlateral hand for patients [36]—to control the motion of robotic finger thus limiting the use of thumb in completion of tasks. Moreover, it offers few user control inputs to control the motion of the robotic finger and force control is not straightforward. In order to overcome these limitations, in particular to obtain multiple user control inputs to control the motion of extra-robotic finger as well as to regulate its compliance, we started exploring EMG signals as possible control interface [2, 109, 142].

The EMG signal is a biomedical signal that measures electrical currents generated in a muscle during its contraction representing neuromuscular activities. The EMG signal is controlled by the nervous system and is dependent on the anatomical and physiological properties of muscles. The main reason for the interest in EMG signal analysis is in clinical and biomedical applications as well as to study the human interactions with the physical world like in case of human-robot interaction. EMG measures the electrical potential between a ground electrode and a sensor electrode. It is possible to measure signals either within the muscle (invasive EMG) or on the skin above a muscle (surface EMG) [143]. The expressiveness of the latter is comparatively limited, but in return, the approach is more easily applicable and does not need special preparation of the skin nor special medical training of the operator. EMG signal acquires noise while traveling through different tissues. Moreover, the EMG detector, particularly if it is at the surface of the skin, collects signals from different motor units at a time which may generate interaction of different signals. Detection of EMG signals with powerful and advance methodologies is becoming a very important requirement in biomedical engineering. The control of the finger flexion/extension through an EMG signal captured by surface electrodes placed on the user's frontalis muscle was presented in [37]. The EMG interface was able to detect if the frontalis muscle of the user was voluntary contracted. The frontalis muscle is always spared in case of stroke due to its bilateral representation in the brain cortex. This makes the frontalis muscle a suitable candidate for the control of an intrinsically compliant underactuated extra finger for grasp compensation as that proposed in [2]. However, when more dexterous robotic devices are considered and when more sophisticated manipulation motions are required, the use of a single muscle is not sufficient to control both motion and force exerted. For that reason, this chapter also presents a novel EMG based control interface which not only can control different trajectories of multi degree of freedom finger but also can regulate its compliance.

## 6.2 The *Frontalis Muscle Cap*: A Wireless EMG System for the Control of Underactuated Soft Supernumerary Robotic Finger

The control interface for patient oriented devices must be intuitive and simple, since chronic patients may also be affected by some cognitive deficits, possibly limiting their compliance during a demanding learning phase. We propose the *frontalis muscle cap*: an Electromyography (EMG) based wireless interface which maintains the principle of simplicity of a switch without interrupting the patient activities and without the involvement of healthy hand during task execution. The *frontalis muscle cap* is a wearable wireless EMG interface where electrodes, acquisition and signal conditioning boards are embedded in a cap, Fig. 6.1. A preliminary version of the control interface has been presented in [2]. This solution allows the patients to autonomously wear the interface using only their healthy hand. Several EMG interfaces have been already successfully adopted for the control of prosthesis [144] and exoskeletons [51]. The electrodes are usually placed either in the muscles coupled with the robot (exoskeleton) or in muscles where amputees still have the phantom of functions and hence they are able to generate a repeatable EMG pattern corresponding to each of the functions (prosthesis). For chronic patients it is generally difficult to generate repeatable EMG patterns in their paretic upper limb due to the weakness in muscle contraction control. For this reason, we coupled the flexion/extension motion of the robotic device with the contraction of the frontalis muscle. This muscle is always spared in case of a motor stroke either of the left or of the right hemisphere

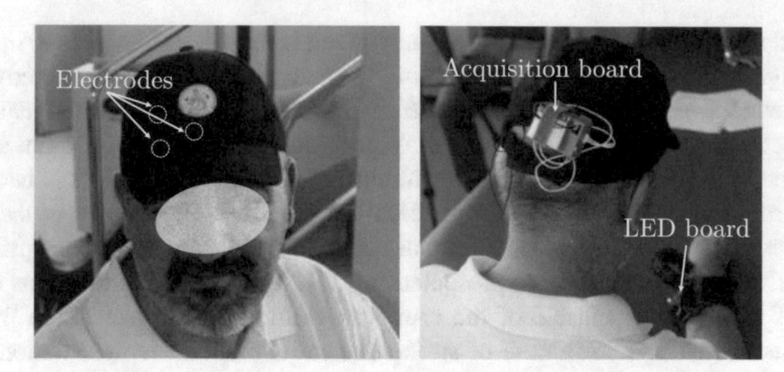


**Fig. 6.1** The *frontalis muscle cap* interface. The EMG electrodes are placed inside the cap at front side to be positioned on the patient's forehead. The acquisition board is placed in a box on the back of the cap

due to its bilateral cortical representation. The user can contract this muscle by moving the eyebrows upwards. The electrodes in the *frontalis muscle cap* capture the arising EMG signal that is acquired through an EMG signal conditioning circuit and processed by a control algorithm as explained in the following.

EMG measures the electrical potential between a ground electrode and a sensor electrode. It is possible to measure signals either within the muscle (invasive EMG) or on the skin above a muscle (surface EMG) [143]. We used surface EMG electrodes to measure electrical signals associated with the patient's frontalis muscle. In particular, on the inner side of the *frontalis muscle cap*, we installed non-gelled reusable silver/silver-chloride electrodes, as they present the lowest noise interface and are recommended for biopotentials recording [145]. We designed an EMG signal acquisition board taking into consideration the requirements associated with bandwidth, dynamic range and physiological principles. The typical EMG waveform is characterized with a spectral content between 10 to 250 Hz with amplitude up to 5 mV, depending on the particular muscle [146].

Figure 6.2 shows the block diagram of the implemented EMG circuit board. Three electrodes are interfaced to the board: two of them ($V_{IN+}$ and $V_{IN-}$) are connected to the inputs of an instrumentation amplifier (In-Amp), while the third one called "ground electrode" is connected to a mid-supply reference voltage ($V_{ss} = V_{cc}/2$). This configuration improves the quality of EMG signal acquisition as it increases the common mode rejection ratio (CMRR). The first stage of the EMG board is an In-Amp with an additional stage of AC-coupling. This configuration allows a precise control of DC levels rejecting undesired DC offset voltage introduced by electrode-skin interface. The DC component is subtracted by feeding the output signal back to the reference input of the In-Amp, by an integrator feedback network, which results in a first-order highpass response. The second stage of the EMG board is a 4th order lowpass Butterworth filter. An active topology (a Sallen-Key circuit implementation −4th order low-pass filter cascading two stages of 2nd order) was chosen to get a better performance and less complexity than a passive one. The specifications of the EMG acquisition board are summarized in Table 6.6.

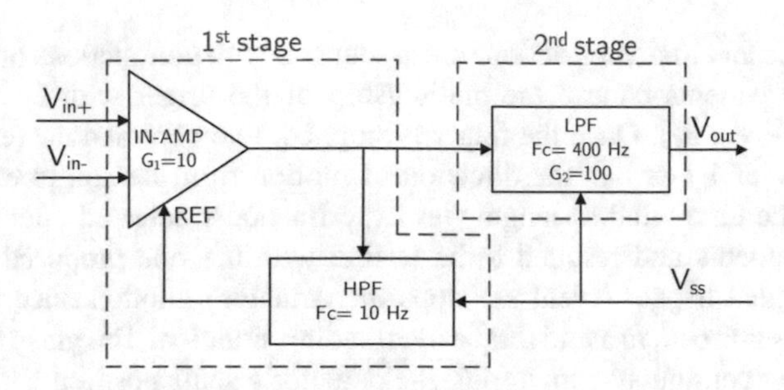


**Fig. 6.2** Block diagram of the EMG circuit board. $V_{IN+}$ and $V_{IN-}$ are the "detecting electrodes" while $V_{ss} = V_{cc}/2$ is the "ground electrode". $REF$ is the reference voltage terminal of the In-Amp

The acquired EMG signal is sampled at 1 kHz (double EMG band) to avoid aliasing and a wireless communication is realized by a pair of Xbee modules (Series 1). The transmitter is embedded in the *frontalis muscle cap* while the receiver is placed on the actuator controller unit. The reference value of received EMG were normalized using maximum voluntary contraction (MVC) technique [147]. This solution avoids the problems related to the high influence of detection condition on EMG signal amplitude. In fact, amplitude can greatly vary between electrode sites, subjects and even day to day measures of the same muscle site. We implemented an auto-tuning procedure based on the MVC in order to better match the user-dependent nature of the EMG signal. This is the major improvement with respect to the *frontalis muscle cap* version proposed in [2] where the sensitivity of the system was manually set through a potentiometer. The implemented MVC routine consists of a 3 s time window in which the users slowly start increasing the contraction of the forehead muscle to reach their maximum effort. The MVC value itself is not calculated as a single peak data point to avoid high variability. In order to obtain a more stable reference value, we have implemented an algorithm using a sliding window technique of 500 ms duration to compute the mean amplitude of the highest signal portion acquired during the 3 s time window.

The motion of compensatory robotic device is then controlled by using a finite state machine based on a trigger signal. The trigger signal is obtained by using a single-threshold value defined as the 50% of the MVC, a level that was repeatable and sustainable for the subject without producing undue fatigue during the use of the device. We set a minimum time (20 ms) in which the EMG signal has to constantly stay over the threshold to generate the trigger signal to prevent false activation due to glitches or to spontaneous spikes. Figure 6.12 shows the raw EMG signal and the signal after the conditioning operations, i.e., rectification, normalization and filtering. The red signal shown in bottom graph of Fig. 6.12 is the resultant trigger signal which is generated if the EMG signal exceeds the threshold.

We implemented a trigger signal based finite state machine (FSM) [148] to control the motion of the devices. The outputs of the FSM are predefined commands based on sequences of input signals. We consider a finite number of states, transition between those states, and commands. States represent predefined motion commands for the robotic device and transition actions are associated with contractions of the frontalis muscle. The patients control the motion/stop of the finger with a single muscle contraction (event $e_1$). Once the finger is stopped, two contractions (event $e_2$) in a time window of 1 s switch the direction of motion from flexion to extension and viceversa. The time window length was experimentally selected after the repeated trials with patients and resulted to be in-line with the one proposed in [149]. A software defined trigger (event $e_4$) stops the actuator's motion once the object is considered as grasped, to avoid torque overloading situation. The grasp confirmation is detected by continuous monitoring the actuator's shaft position and the exerted torque. During the grasping procedure, if the position does not change in a time window of 2 s and a predefined torque threshold is reached, the object is considered as grasped. The proposed FSM is reported in Fig. 6.14 (Table 6.1).

A LED board (see Fig. 6.3) is used to provide a visual feedback of the selected commands. In particular, a yellow LED blinks on each trigger signal. When flexion is selected an orange LED is turn on, while a green LED shows the extension. Finally a red LED is turn on when the device is stopped. At this stage, the LED associated to the previous selected state is also turned on to remind the user about the last stage of the device (Fig. 6.4).

To provide an additional interface for the user, as well as a recovery mode for possible problem in the *frontalis muscle cap* communication, we added a push-button on the LED board as further possible control. Both interfaces use the same trigger-based logical scheme to control the motion of robotic devices. The switching between two control interfaces can be achieved at any moment by a toggle switch installed on the controller board. If the *frontalis muscle cap* is selected, the procedure of maximum voluntary contraction (MVC) is first executed. Once it is completed, the program passes from calibration to *test mode*. If the push-button is selected, the control algorithm directly jump to the *test mode*. In *test mode*, the user tests the selected interface by displaying the output on the LEDs mounted on the LED board to get familiar with the interface without using the robotic device. When the *frontalis muscle cap* control interface is selected, the user can do the MVC calibration again (if needed) by simply pressing once the push-button. If the *test mode* output confirms

**Table 6.1** Specifications of EMG acquisition board

| | |
|---|---|
| EMG acquisition box dimensions | $35 \times 31 \times 45$ mm$^3$ |
| EMG acquisition box weight | 46 g |
| Principle | Differential voltage |
| Number of electrodes | 3 |
| Bandwidth | 10–400 Hz |
| Gain | 1000 |
| Input impedance | 100 GΩ |
| CMRR | 110 dB |
| Operating voltage | $Vcc = 3.3$ V |

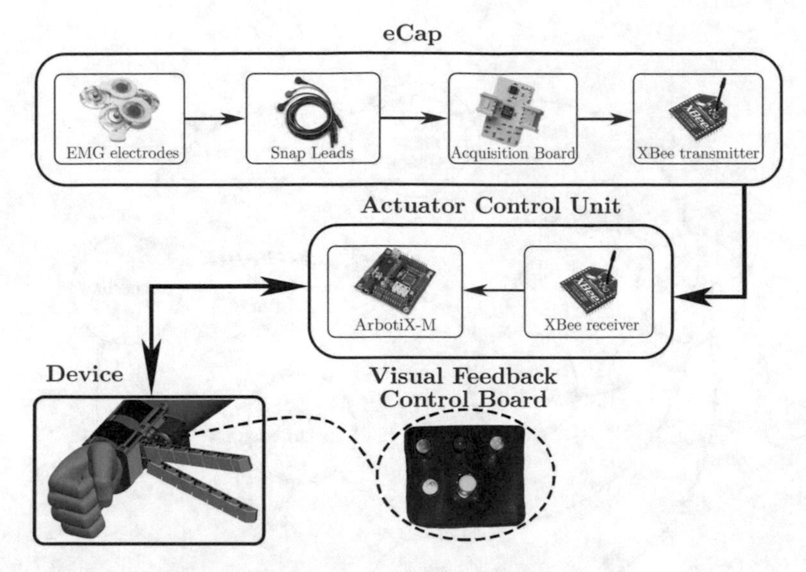


**Fig. 6.3** The block diagram of EMG wireless and push-button interface with actuator control unit. LEDs associated to the motion of device are mounted on visual feedback control board

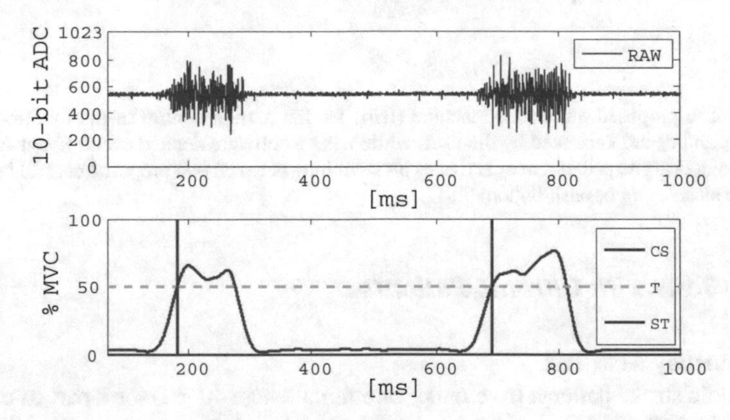


**Fig. 6.4** Top: raw EMG signal. Bottom: example of two activations in a time window of 1 s. CS (blue) is the processed EMG signal after the operations of rectification, normalization and filtering; T (green) is the threshold and ST (red) is the resulting trigger

to the expected program behavior and user is ready to use the device, he or she can switch to the *device mode*, where, motion of robotic device is controlled according to FSM in Fig. 6.14. The switching from *test mode* to *device mode* is achieved by pressing the push-button for more than one second and is represented in the FSM in Fig. 6.14 by the event $e_5$. The process repeats every time the user switch to other control interface (Fig. 6.5).

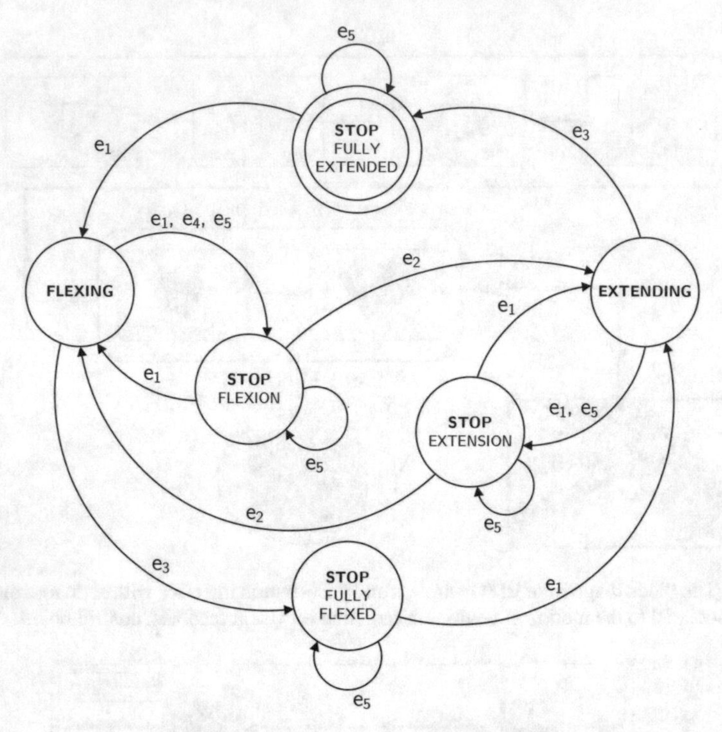


**Fig. 6.5** The proposed finite state machine (fsm) for the motion control of the robotic devices. Events $e_1$ and $e_2$ are generated by the user, while $e_3$ is a software defined event. Event $e_4$ occurs once the object is grasped. Event $e_5$ activates on switching between two proposed control interfaces (*frontalis muscle cap* or push-button)

## 6.2.1 Tests with Chronic Patients

**The Frenchay Arm Test**

Six chronic stroke patients (five male, one female, age 40–62) took part to the tests on how the Soft-SixthFinger can be used for hand grasping compensation. The proposed compensatory tool can be used by subjects showing a residual mobility of the arm. For being included in the experimental phase, patients had to score $\leq 2$ when their motor function was tested with the National Institute of Health Stroke Scale (NIHSS) [84], item 5 "paretic arm". Moreover, the patients had to show the following characteristics: normal consciousness (NIHSS, item 1a, 1b, 1c = 0), absence of conjugate eyes deviation (NIHSS, item 2 = 0), absence of complete hemianopia (NIHSS, item 3 $\leq$ 1), absence of ataxia (NIHSS, item 7 = 0), absence of completely sensory loss (NIHSS, item 8 $\leq$ 1), absence of aphasia (NIHSS, item 9 = 0), absence of profound extinction and inattention (NIHSS, item 11 $\leq$ 1).

The goal of the tests was to verify how quickly the patients can learn to use the device and its EMG control interface. We performed a fully ecological qualitative test, the Frenchay Arm Test [81]. The test consisted of five tasks to be executed within three minutes:

1. *Task 1*: Stabilize a ruler, while drawing a line with a pencil held in the other hand. To pass, the ruler must be held firmly.
2. *Task 2*: Grasp a cylinder (12 mm diameter, 5 cm long), set on its side approximately 15 cm from the table edge, lift it about 30 cm and replace without dropping.
3. *Task 3*: Pick up a glass, half full of water positioned about 15–30 cm from the edge of the table, drink some water and replace without spilling.[1]
4. *Task 4*: Remove and replace a sprung clothes peg from a 10 mm diameter dowel, 15 cm long set in a 10 cm base, 15–30 cm from table edge. Not to drop peg or knock dowel over.
5. *Task 5*: Comb hair (or imitate); must comb across top, down the back and down each side of head.

The patient scored 1 for each of the successfully completed task, while he or she scored 0 in case of fail. The subject sat at a table with his hands in his lap, and each task started from this position. He or she was then asked to use his or her affected arm/hand to execute the tasks. Although the Frenchay arm test has not been specifically designed for evaluating compensatory tools, it has shown good reliability in measuring functional changes in stroke patients when comparing with other upper limb assessments [81]. The Frenchay Arm Test contains manipulation tasks of everyday life activity that evaluate the capability of grasping an object without a deep involvement of the patient arm. Other tests like, e.g., ARAT and UEFT [150], are more suitable for the evaluation of arm mobility which is out of the scope of this work.

Patients wore the Soft-SixthFinger in their paretic limb, the left hand for two subjects and the right one for the other four. The number of modules on the device was chosen according to the selected tasks. Seven modules were used in order to grasp the proposed objects.Written informed consent was obtained from all participants. The procedures were in accordance with the Declaration of Helsinki. The rehabilitation team assisted the subjects during a training phase that lasted for about one hour. During this phase, the optimal position of the device on the arm, according to the patient motor deficit, was evaluated. Moreover, the *frontalis muscle cap* parameters were tuned according to the patient. After the training phase, the subjects had three minutes to perform the Frenchay Arm Test. Three patients tried the extra-finger for the first time. All the subjects performed the test two times, one without and one with the Soft-SixthFinger. The starting condition was selected randomly. The results of the test are shown in Tables 6.2 and 6.3 for the six patients. Screenshots of the tasks are reported in Fig. 6.6. Note that in the execution of Task 1 the Soft-SixthFinger do not interfere with the paretic limb action. The patients can stabilize the ruler without using any external help, so the device is kept in its rest position.

[1]Note that for safety reasons we did not use water in presence of electronic components.

**Table 6.2** Results of the Frenchay Arm Test for the patient 1 to patient 3 with and without using the Soft-SixthFinger (SSF)

| Frenchay Arm Test | Patient 1 | | Patient 2 | | Patient 3 | |
|---|---|---|---|---|---|---|
| | With | Without | With | Without | With | Without |
| Stabilize a ruler | 1 | 1 | 1 | 1 | 1 | 1 |
| Grasp a cylinder | 1 | 0 | 1 | 0 | 1 | 0 |
| Pick up a glass | 1 | 0 | 1 | 0 | 1 | 0 |
| Remove a sprung | 0 | 0 | 0 | 0 | 0 | 0 |
| Comb hair | 0 | 0 | 0 | 0 | 0 | 0 |

**Table 6.3** Results of the Frenchay Arm Test the patient 4 to patient 6 with and without using the Soft-SixthFinger (SSF)

| Frenchay Arm Test | Patient 4 | | Patient 5 | | Patient 6 | |
|---|---|---|---|---|---|---|
| | With | Without | With | Without | With | Without |
| Stabilize a ruler | 1 | 1 | 1 | 1 | 1 | 1 |
| Grasp a cylinder | 1 | 0 | 1 | 0 | 1 | 0 |
| Pick up a glass | 1 | 0 | 1 | 0 | 1 | 0 |
| Remove a sprung | 0 | 0 | 0 | 0 | 0 | 0 |
| Comb hair | 0 | 0 | 0 | 0 | 1 | 0 |

### Bimanual Tasks

The latter phase of post-stroke rehabilitation is based on the learning of newly acquired motor strategies to compensate the neurological deficit. These strategies may sometimes be neither ergonomic nor ecological, or may even increase pathological motor patterns, usually by worsening tonic flexion at the forearm of the paretic limb [151]. Some compensation techniques take also advantage of dedicated objects that allows the users to execute typical bimanual tasks only with the healthy hand. An example of available aids is reported in Fig. 6.7. However, the use of such tools is usually limited to the structured houses of the patients, restricting the possibilities of the patients to exploit them outside. The Soft-SixthFinger is a portable compensatory tool that can be carried as a bracelet when not used. This allows the patients to bring the devices wherever they want. The regained capability of grasping object with the help of the device, stimulate the patient to use his paretic limb so preserving residual

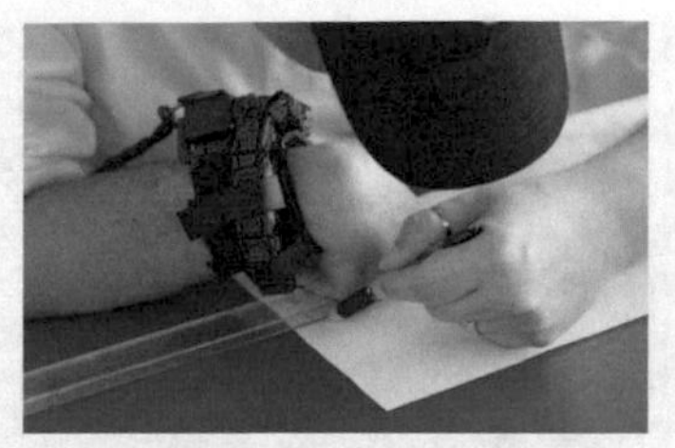

Task 1: Stabilize a ruler

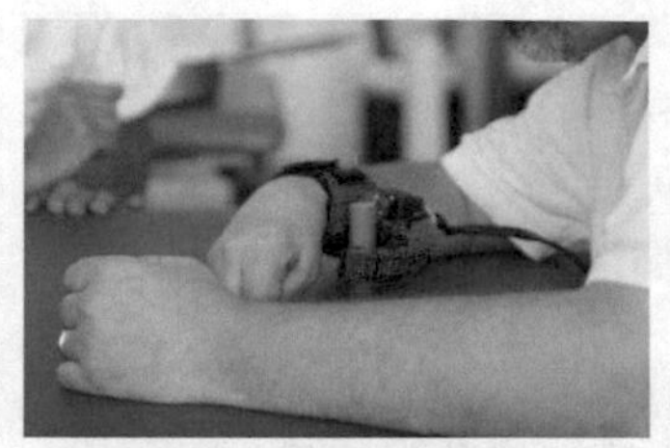

Task 2: Grasp a cylinder

Task 3: Pick up a glass

Task 4: Remove a sprung

Task 5: Comb hair

**Fig. 6.6** The five tasks of the Frenchay Arm Test

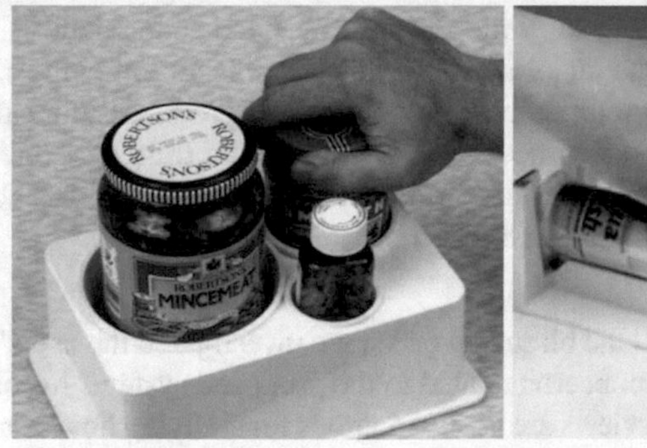


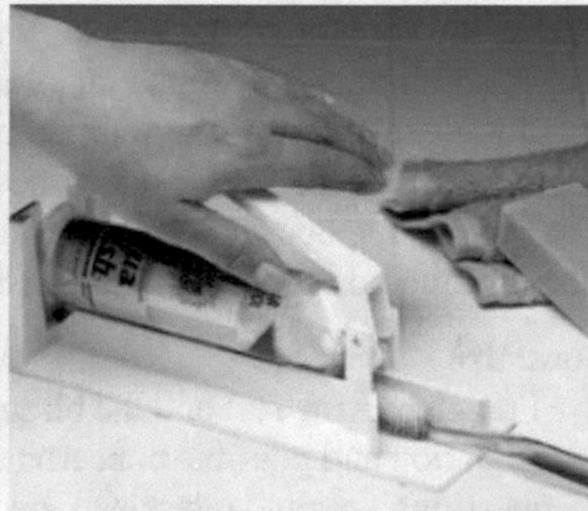

**Fig. 6.7** Examples of adaptive kitchen aids designed to be used with only one hand. On the left, a case used as aid for bottle and jar opening. On the right, a toothpaste dispenser. Pictures courtesy of Elderstore®

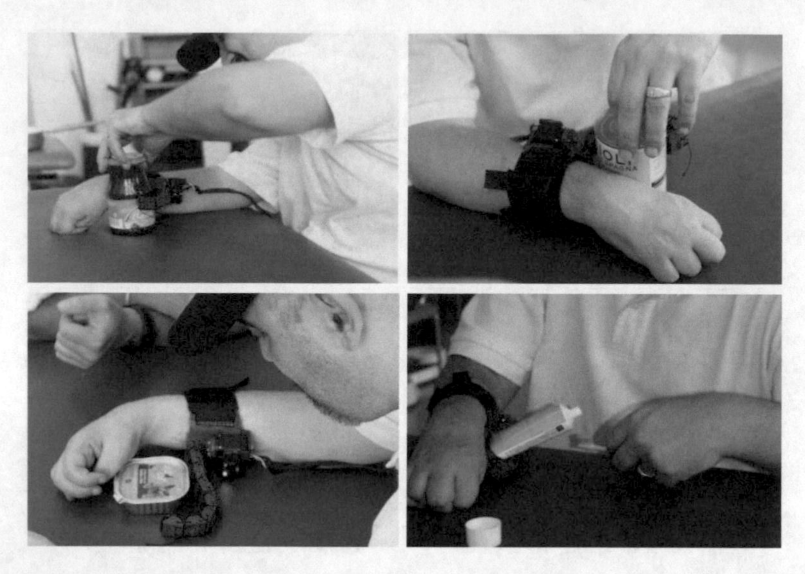

**Fig. 6.8** The four tasks. From the top left, clockwise: unscrew a jar, open a can, squeeze a toothpaste tube and open a squared can

mobility. As a proof of concept, we tested the device in four different bimanual tasks typical of ADL. The four tasks were:

- unscrew the cap of a tomato jar;
- open a can of beans;
- open a squared can of cat food;
- squeeze toothpaste tube over a toothbrush.

In all these bimanual tasks, the paretic limb and the Soft-SixthFinger work together to constrain the motion of the object while the healthy hand manipulate it (e.g., constrain the motion of the tomato jar while the healthy hand is unscrewing its cap). Pictures of the execution of the four tasks are reported in Fig. 6.8. All the patients were able to execute the four tasks without requiring a specific training. Note that, thanks to the passive compliance embedded in its structure, the device was able to adapt also to squared shapes like the can of cat food.

**Questionnaire**
After the Frenchay Arm Test and the bimanual tasks, we investigated the users' subjective satisfaction and possible concerns related to the proposed system. According to [152], questionnaires and interviews are useful methods for studying how users use systems and what features they particularly like or dislike. The patients were asked to fill the Usefulness-Satisfaction-and-Ease-of-use-questionnaire (USE, [124]) that focuses on the experience of the system usage. This questionnaire uses a seven-point Likert rating scale. Mean and standard deviation (SD) of the questionnaire factors are presented in Table 6.4.

**Table 6.4** Questionnaire factors and relative marks. The mark ranges from "1 = strongly disagree" to "7 = strongly agree". Mean and standard deviation (Mean (SD)) are reported

| Questionnaire factors | Mean (SD) |
|---|---|
| Usefulness | 4.6(0.8) |
| Ease of use | 5.6(0.6) |
| Ease of learning | 6.5(0.8) |
| Satisfaction | 5.8(0.7) |

## 6.3 An EMG Interface for the Control of Motion and Compliance of Fully Actuated Supernumerary Robotic Finger

In this part of the chapter, we present an EMG interface for the control of motion and compliance of fully actuated supernumerary robotic finger. The motion of the robotic extra finger is controlled through gesture recognition and its compliance is regulated by EMG signal amplitude variations. In particular, we used Myo Armband for gesture recognition to be associated with the motion control of the robotic device and surface one channel EMG electrodes interface to regulate the compliance of the robotic device.

The study also present an updated version of the fully actuated robotic extra finger where the adduction/abduction motion is realized through ball bearing and spur gears mechanism. The proposed interface is validated with two sets of experiments related to compensation and augmentation. In the first set of experiments, different bi-manual tasks have been performed with the help of the robotic device and simulating a paretic hand since this novel wearable system can be used to compensate the missing grasping abilities in chronic stroke patients. In the second set, the robotic extra finger is used to enlarge the workspace and manipulation capability of healthy hands. In both sets, the same EMG control interface has been used. The obtained results demonstrate that the proposed control interface is intuitive and can successfully be used, not only to control the motion of a supernumerary robotic finger, but also to regulate its compliance. The proposed approach can be exploited also for the control of different wearable devices that has to actively cooperate with the human limbs.

### *6.3.1 The Fully Actuated Supernumerary Robotic Finger*

The proposed supernumerary robotic finger is composed of modules connected to partially resemble the human finger mechanical structure. Human hand fingers, excluding the thumb, consists of four phalanges connected by three joints [62]. The structure of the thumb is different since it has two joints at the base for the anter-position or retroposition combined with the radial or palmar abduction motions.

The other fingers are capable of both adduction-abduction and flexion-extension motions. The finger's kinematic model is typically approximated by using simple revolute joints. This approximation is an effective means of modeling, as these are in fact the same as compared to proximal and distal joints of humans. The proximal and distal interphalangeal articulations can have only flexion/extension motion capabilities and typically are represented with a single DoF revolute joint. The metacarpal joints have both adduction/abduction and flexion/extension motion capabilities and can be modeled as a 2-DoFs joint composed of two revolute joints with orthogonal rotation axis (universal joint).

We designed the kinematic structure of the robotic extra finger such that one motor is adopted to actuate each DoF of the robotic finger so to replicate the flexion/extension motion of the human finger. While, at the robotic finger base, two motors realize the adduction/abduction and flexion extension motion to replicate metacarpal joint. We used four modules in a pitch-pitch configuration for the flexion/extension motion of the finger so to approximate the average length of the whole hand [153]. The adduction/abduction motion of base joint is obtained using spur gears that allows to transmit motion and power. One of the spur gear is mounted on the shaft of the servo motor, while the other is placed on the base of the finger. We used bearings to decrease the friction during rotation.

The finger design is based on the principle of modularity. Each module consists of a servomotor, a 3D printed structure (Acrylonitrile Butadiene Styrene, ABSPlus, Stratasys, USA) and a soft rubber part mounted on front to increase the friction at the contact area. The actuators used are the HS55 MicroLite servo motors. The modules are connected so that one extremity of each module is rigidly coupled with the shaft of the motor through screws, while the other has a pin joint acting as revolute joint. The exploded view and the prototype of the device are shown in Fig. 6.9.

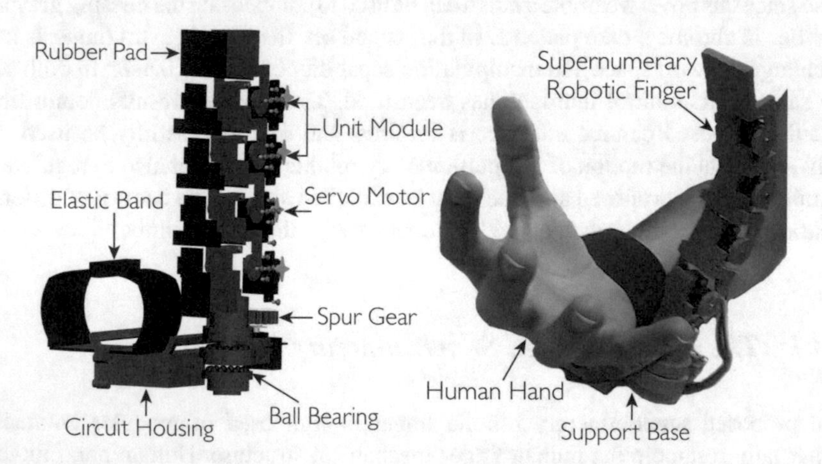


**Fig. 6.9** On left the exploded cad view, while on right the prototype of the robotic extra finger. Four modules are used for the flexion/extension motion, while the revolute joint based on bearings and spur gears mechanism at the finger base is used for the adduction/abduction motion. The device can be worn on the forearm through an elastic band

**Table 6.5** The technical details of supernumerary robotic finger

| | |
|---|---|
| Device weight | 0.16 kg |
| Module dimension | 42 × 33 × 20 mm |
| Module weight | 16 g |
| Support base dimension | 78 × 24 × 5 mm |
| Support base weight | 28 g |
| Max torque per motor | 0.15 Nm |
| Max payload | 0.61 kg |
| Velocity of one module | 0.5 rad/s |
| External battery pack | 5 V |

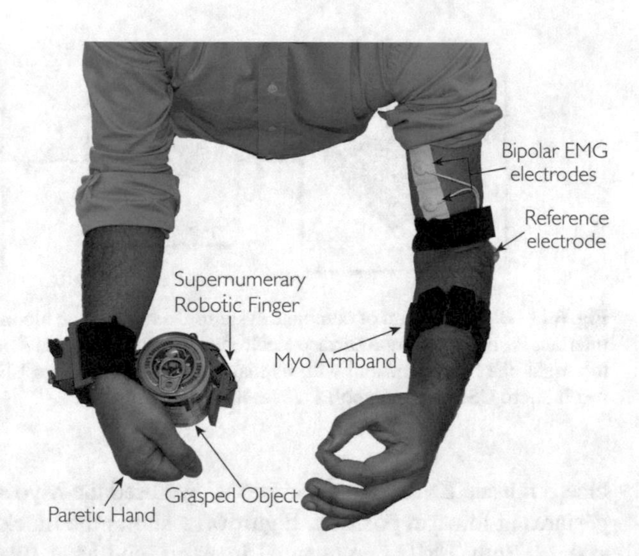


**Fig. 6.10** The complete system: the EMG interface on one arm while the supernumerary robotic finger on the other arm. Myo Armband is positioned on the forearm while the one channel interface is placed on the biceps muscle

The servo motors are pulse width modulation (PWM) controlled. The PWM signals are generated by a microcontroller At-mega 328 installed on an Arduino Nano board. The portability and wearability of the device is improved by enclosing all the electronics circuitry in a 3D printed housing which is attached to the finger base support. An external battery pack (5 V) is used to provide power to the actuators. Technical details on the device are summarized in Table 6.5.

We used surface EMG electrodes to measure electrical signals associated with the subject muscle activation. In particular, we used two EMG interfaces on the arm, one to record the continuous EMG amplitude aiming to regulate the compliance of the device and the second to recognize different gestures to be associated with the motion of the robotic finger. Both EMG interfaces are placed on one of the arms, one at the biceps and other at the forearm, while having the robotic finger on other arm as shown in Fig. 6.10. We developed the circuit acquisition and signal conditioning board for one channel EMG electrodes to measure continuously the

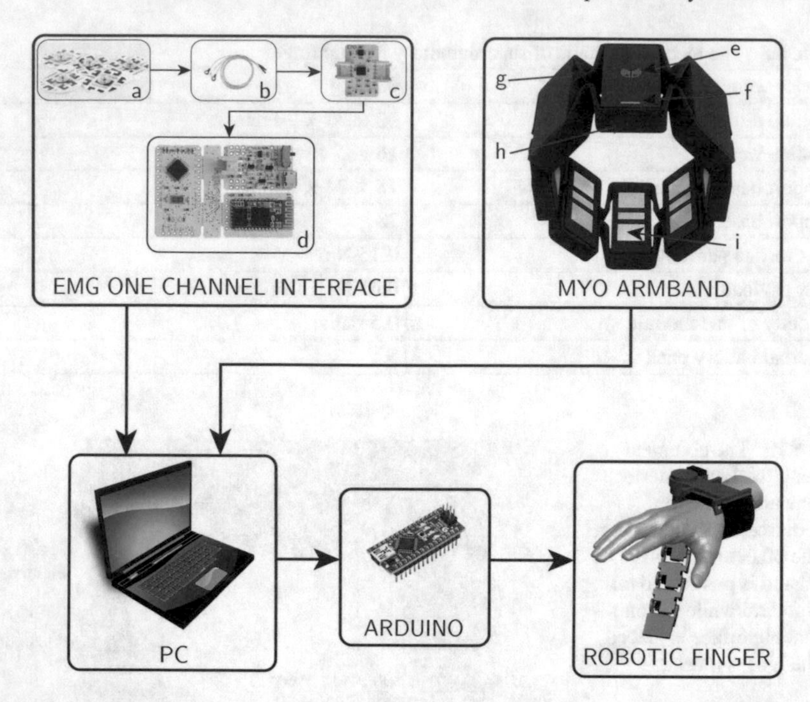


**Fig. 6.11** Block diagram of complete system. On top left, the block diagram of EMG one channel interface is shown, where **a** surface electrodes **b** snap leads **c** acquisition board **d** control board. On top, right, the Myo Armband with its major components **e** logo LED **f** status LED **g** expandable flex **h** micro USB charging port **i** electrical sensor

biceps muscle EMG signal variations. We used the Myo Armband to recognize the gestures at forearm position. Figure 6.11 shows the block diagram of the proposed system. Both, EMG one channel interface and Myo Armband are connected to a computer through Bluetooth communication. The PC communicates with the robotic device controller (Arduino) through serial communication which in turn controls the motion and compliance of the supernumerary robotic finger. Section 6.3.2 describes the development of the acquisition and signal conditioning board for one channel EMG interface followed by the compliance regulation of the robotic device through the amplitude variation of the acquired biceps EMG signal. We also describe the gesture recognition through the Myo Armband and their association with the motion control of the supernumerary robotic finger through a finite state machine (FSM).

### 6.3.2 One Channel EMG Electrodes Interface and Robotic Device Compliance Regulation

We used non-gelled reusable silver/silver-chloride electrodes for the EMG one channel interface. These are recommended for biopotentials recording since they present the lowest noise interface [145]. The design and development of the EMG signal acquisition board is carried out, while considering the requirements associated with bandwidth, dynamic range and physiological principles. The typical EMG waveform is characterized with a spectral content between 10 to 250 Hz with amplitude up to 5 mV, depending on the particular muscle [146]. The acquired EMG signal is sampled at 1 kHz (double EMG band) to avoid aliasing. The reference value of received EMG were normalized using maximum voluntary contraction (MVC) technique [147]. This solution avoids the problems related to the high influence of detection condition on EMG signal amplitude. In fact, amplitude can greatly vary between electrode sites, subjects and even day to day measures of the same muscle site. We implemented an auto-tuning procedure based on the MVC in order to better match the user-dependent nature of the EMG signal. The implemented MVC routine consists of a 3 s time window in which the user slowly starts increasing the contraction of the biceps muscle to reach their maximum effort. Figure 6.12a shows the relation between EMG (percentage of MVC) signal at biceps and time (ms). The relationship between EMG signal and muscle tension is non linear. The MVC value itself is not calculated as a single peak data point because that would allow too much variability. In order to obtain a more stable reference value, we have implemented an algorithm using a sliding window technique of 500 ms duration to compute the mean amplitude of the highest signal portion acquired during the 3 s time window. The technical details of the EMG acquisition board are listed in Table 6.6. The EMG signal acquired through the developed one channel interface is used to control the stiffness of each module of the robotic device through the implemented control scheme based on servo motor, explained later in this section. In precision grasps, the contact is obtained at the fingertip module. In power grasps, in order to obtain suitable contact points, we set different closing priorities according to the position of the module in the finger. If the fingertip module comes in contact first, the remaining modules stop. If another module comes in contact first, modules below to it stop, while the module above keeps moving. The same methodology is followed for other intermediate modules. Finally, if the module at the base of the finger is the first to come in contact, two different behaviors can occur: (i) one of the intermediate modules get in contact before the fingertip; (ii) the fingertip module comes in contact first. In case (i) the modules above are left free to move to get in contact with the object; in case (ii) the other intermediate modules stop. The contact of a module can be detected comparing the desired angle commanded to the servo motor ($q_{des}$) with the actual position read by the encoders ($q_m$). When $\Delta q = \|q_{des} - q_m\|$ overtake a predefined threshold a contact is recognized. It is possible to control the compliance of the device to guarantee the stability of the grasp. During the grasping phase the finger should be compliant so to favor the adaptation of the finger to the shape of

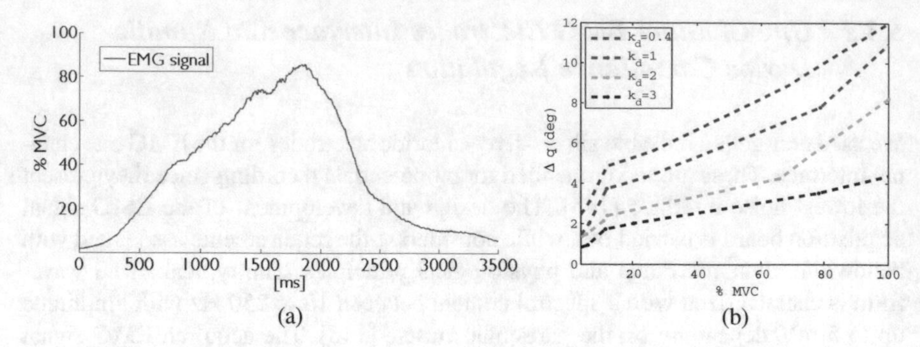


**Fig. 6.12** On left, **a** the maximum voluntary contraction (MVC) proportional to the biceps muscle contraction is shown. While on right, **b** the graph between $\Delta q$ and percentage of MVC for different values of $k_d$ is plotted

**Table 6.6** Technical details of EMG signal acquisition and conditioning board

| | |
|---|---|
| EMG acquisition box dimensions | $3.5 \times 3.1 \times 4.5$ cm$^3$ |
| EMG acquisition box weight | 46 g |
| Principle | Differential voltage |
| Number of electrodes | 3 |
| Bandwidth | 10–400 Hz |
| Gain | 1000 |
| Input impedance | 100 GOhm |
| CMRR | 110 dB |
| Operating voltage | $Vcc = 3.3$ V |

the grasped object. Once the grasp is achieved, the stiffness of the finger should be increased so to tight the grasp. The finger stiffness is regulated from the user through the EMG interface by contracting the biceps. This approach is similar to the concept of teleimpedance introduced by Ajoudani et al. [154].

In the following we will explain how stiffness regulation has been obtained using servomotors. Generally, in active compliance control framework, the equation relating the motor torque to its position is given by

$$\tau = k\Delta q = k(q_{des} - q_m) \tag{6.1}$$

where, $q_{des}$ is the desired (reference) joint position , $q_m$ is the measured (current) joint value, and $k$ is the stiffness constant [83]. Note that the compliant (or stiff) behavior of the joint is achieved by virtue of the control, differently from what happen in mechanical systems with a prevalent dynamics of elastic type.This controller is typically used with actuator that can be torque controlled. Servo motors are position controlled actuators where it is not possible to directly command the exerted torque. Even if the joint position deviates a little from the reference position in the clockwise direction, a large amount of torque is generated in the counterclockwise direction to compensate for this. This behavior is regulated by the controller embedded in the servo motor and cannot be modified. However, since inertia must be considered, we set up a rack-pinion gear mechanism connected with springs with calibrated stiffness to estimate the real relation between the angle error (difference between commanded and actual position of the motor) and the exerted torque. We called the arising stiffness constant of the motor $k$. We modified the servo motor to read its current position. The pinion gear was mounted on the shaft of the motor. One end of the rack gear was coupled with the motor pinion gear, while the other was attached to two springs with known stiffness. The other end of the springs were fixed (grounded). Once the servo moves to its goal position, it applies the force on the springs through the rack-pinion gear arrangement. The corresponding motor torque can be computed by using the relationship $\tau_m = F\,r$ where, $r$ is the radius of the gear and $F$ the force applied on the springs. $F$ can be computed using $F = k_{sp}\,x$. Here, the spring constant $k_{sp}$, is known and displacement $x$, is measured through a scale. This torque-position relationship defines the standard stiffness of the servo motor which can not be changed by the user. Thus, the ultimate choice to vary the torque is the position error (second term in Eq. 6.1). We defined a scaling factor $k_d$ to modulate the position error.

Thus the possible angle displacement $\Delta q_i(t)$ at time instant $t$ for the $i$th servo motor was computed as

$$\Delta q_i(t) = k_d(q_{des} - q_m),$$

We modified the servomotor in order to measure its actual position by accessing the internal encoder measurements. The only servomotor parameter that can be commanded is its desired position $q_{des}$. At time instant $t$ the desired position for the $i$th servomotor is computed as

$$q_{des,i}(t) = q_{m,i}(t-1) + \Delta q_i(t-1).$$

In presence of a rigid grasped object, the measured positions of the extra-finger joints do not change due to the object constraints. So that, changing the desired position of the servomotors through the scaling factor we can control the force exerted by the device onto the object. For calibration purpose, We mounted a Force Sensing Resistor (FSR) (408, Interlink Electronics Inc., USA) on single module. We pushed the module and recoded the $\Delta q$ against the applied pushing force for different values of $k_d$. We used piecewise-linear functions to replace the force with the EMG signal amplitude variations in the resultant plot. In Fig. 6.12b, the relation between the percentage of MVC and the arising angle displacement for one module is reported for different value of $k_d$. Increasing the value of $k_d$ we get a less angle displacement of the servomotor. Finally, the resulting behavior was implemented in the device controller where the paramcter $k_d$, was varied proportional to the EMG signal acquired at the user biceps. In particular, the range of EMG signals was linearly mapped in the range 0.4–3 of parameter $k_d$.

We set priorities between modules for the simulated compliance variation, similarly to what we did for the grasping procedure. So that, if, for instance, only the

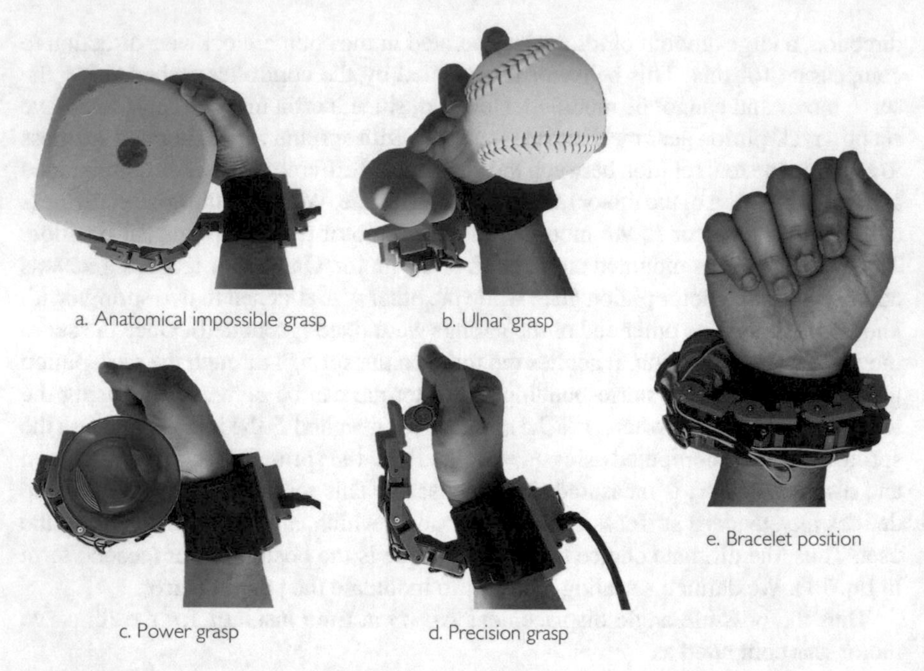


**Fig. 6.13** Examples of possible achievable grasps at working positions (**a**–**d**) and bracelet at rest position (**e**). In **a** and **b** the robotic finger coordinates with healthy hand to realize the *anatomically impossible* and *ulnar* grasp, respectively. While in **c** and **d** it interacts with paretic hand to realize *power* and *precision grasp*

fingertip module is in contact with the object, all the other modules change their stiffness accordingly. This solution allows to control the stiffness of modules that are not in contact with the object.

**EMG Armband Gesture Recognition and Robotic Device Motion Control**
We used a Myo Armband at forearm to recognize the hand gestures that control the device motions. This device has electrically safe setup with low voltage battery and Bluetooth LE protocol, eight surface EMG sensors working at frequency of 2200 Hz and 9-DoF IMU working at 50 Hz. The provided software development kit (SDK) is suitable for working with the recorded data and for developing standalone applications. EMG signals are filtered through notch filters at frequencies of 50 Hz and 60 Hz in order to take out any power-line interference. For the sake of simplicity, we considered the five gestures available with the SDK. These gestures mainly involve flexion/extension of fingers and flexion/extension of hand.

We implemented specific types of grasps for both kinds of users in order to make better suitable to use the robotic finger with healthy hand or paretic hand. In particular, in case of healthy hand, we defined *anatomically impossible* grasps and *ulnar* grasps, see Fig. 6.13a, b. In case of *anatomically impossible* grasp, the supernumerary robotic finger coordinates with human hand to grasp big size objects which can not be grasped using only one hand. In *ulnar* grasp configuration, the robotic device coordinates

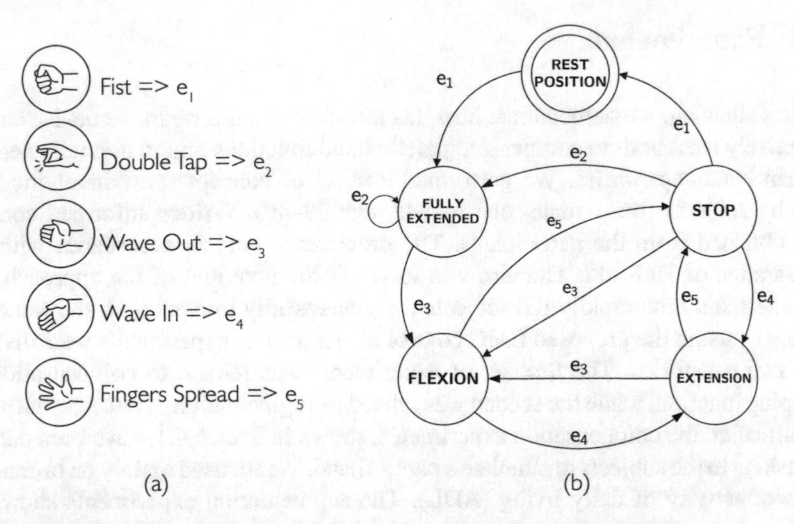


**Fig. 6.14** **a** The recognized gestures and associated trigger signal. **b** The finite state machine which controls the motion of the robotic device in correspond to the generated gesture

with ring and pinkie fingers to grasp and hold an object, while the upper part of the hand (thumb, index and medium fingers) is left free to do another task allowing, for instance, to hold multiple object in one hand or to unscrew a bottle cap with a single hand. In case of paretic hand users, we defined *power* and *precision* grasp as shown in Fig. 6.13c, d. In the former, each module flexes with a fixed step size in order to wrap the finger around the object. In the latter, the target is to hold small size objects between the paretic limb and the device fingertip pad. To this aim, the fingertip is kept parallel to the paretic limb during flexion motion. The contact is expected to occur between the object and the fingertip module. Finally, the supernumerary robotic finger can actively be wrapped around the wrist as a bracelet when not used (see, Fig. 6.13e). We implemented a trigger based FSM to control the motion of the robotic device. All the gestures were associated to a unique trigger signal. In Fig. 6.14a, the gesture recognized through the Myo Armband are shown. In particular, *fist*-(event $e_1$) switches the device from bracelet position to working position and vice-versa. *Double tap*-(event $e_2$) changes the grasp modalities. Patients with paretic hand can switch between precision and power grasp. When augmentation purpose is concerned, the user can switch between ulnar and anatomically impossible grasp. *Wave out*-(event $e_3$) corresponds to flexion and *Wave in*-(event $e_4$) is associated to extension. Finally, *Finger spread*-(event $e_5$) can stop the motion of the robotic finger.

## 6.4 Experiments

In the following, we demonstrate how this interface and the wearable device can be effectively used both to compensate paretic hand functions and to augment healthy human hand capabilities. We performed a proof of concept study involving four healthy subjects (three male, one female, age 29–40). Written informed consent was obtained from the participants. The procedures were in accordance with the Declaration of Helsinki. The aim was to verify the potential of the approach and to understand how rapidly the subjects can successfully interact with the wearable device by using the proposed EMG control interface. The experiments were divided into two categories. The first set of experiments was related to compensation of grasping function, while the second was related to augmentation of hand capabilities. In particular, the compensation experiments, shown in Sect. 6.4.1, have been carried out asking to the subjects to simulate a paretic hand. We focused mainly on bi-manual tasks of activity of daily living (ADL). The augmentation experiments shown in Sect. 6.4.2 were performed with the healthy hand to show the effectiveness of the device in increasing the hand grasping abilities and workspace, e.g., allowing to grasp big size objects which can not be grasped using a single hand or holding multiple objects using the augmented hand, i.e., human hand and the supernumerary robotic finger. In both the experimental sets, the subjects used the EMG interface on one arm (three subject used the right arm, one the left), while the supernumerary robotic finger was worn on the other arm. The Myo Armband was positioned at the forearm, while the one channel electrodes interface on the biceps, see Fig. 6.10.

### *6.4.1 Compensation of Paretic Hand Functions*

Among the different ADL we focused on those involving "hold and manipulate" tasks. Such activities, are generally bimanual tasks where one hand is used to restrain the motion of one object, while the other operates on it, e.g., unscrew the cap of a bottle, open a beans can, etc. The proposed supernumerary finger can be an effective aid in such tasks [2]. To demonstrate how the EMG interface can be used by patients, we asked to the subjects to execute different ADL involving a hold and manipulate task, see Fig. 6.15. In particular, the subject were asked to grasp an object using the gestures of the hand and to regulate the grasp tightness acting on the stiffness of the device. We used a subset of objects from the YCB grasping toolkit [122]. This toolkit is intended to be used to facilitate benchmarking in prosthetic design, rehabilitation research and robotic manipulation. The objects in the set are designed to cover a wide range of aspects of the manipulation problem. It includes objects of daily life with different shapes, sizes, textures, weight and rigidity. We considered six objects with different shapes to show how the robotic finger can adapt to the shape of the objects to realize a stable grasp. We mainly targeted the objects used in kitchen and

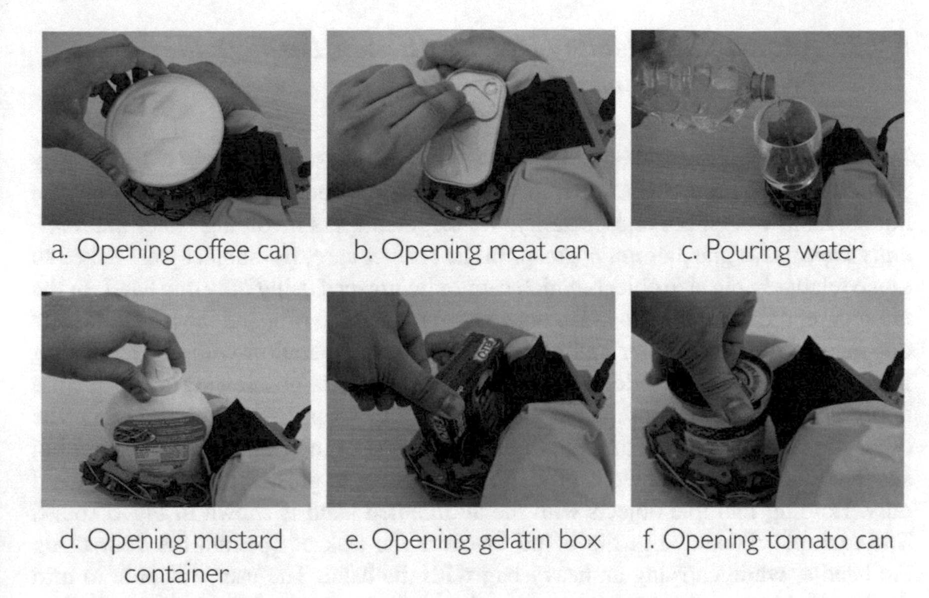


**Fig. 6.15** Supernumerary robotic finger helping in bi-manual task of ADL. All the bi-manual tasks can be completed in the presence of robotic device even if one hand is non-functional

in other ADL. During all the tests, subjects simulated the paretic hand and device was positioned on the arm as supposed to be used with the patients.

The subject was asked to perform different bi-manual ADL without using the hand grasping ability where the device was worn. The controlateral arm was always used to control the device motion and joints stiffness. Figure 6.15 shows the ADL tasks performed by simulating a paretic hand. All the targeted tasks normally require two healthy hands, but have been successfully executed with the aid of the robotic extra finger even if one hand was non-functional. The robotic finger and paretic hand was used to constrain the object, while healthy hand was used for manipulation. Figure 6.15a, b, d, e, f show the example of opening the cans, box and bottle with various shapes and different caps. Figure 6.15c reports the task of pouring water from a bottle while holding the glass with the help of robotic device and paretic arm. All the task were fulfilled controlling the device through the proposes interface. The subjects used hand gestures to shape the finger around the object. Later, they controlled the grasp stiffness by contracting the controlateral arm biceps. Note that all the "opening" tasks required stiffness control to be executed. In fact, while compliant joints are preferable to adapt the shape of the finger to the object to grasp, a stiff device is necessary to achieve the stable grasps necessary while unscrewing the caps.

### 6.4.2 Augmenting Healthy Hand Function Through the Proposed System

In this experiments, the subject were asked to grasp a set of objects with the augmented hand to prove the effectiveness of the extra robotic finger in enlarging the human hand workspace and dexterity. We targeted tasks involving either *anatomically impossible* grasp or *ulnar* grasp. In the former case, the subject were asked to grasp relatively big size objects which cannot be grasped using only one hand. In the latter case, the users tried to grasp objects only using the ring and the pinkie fingers opposite to the sixth finger and to perform another operation with the remaining fingers (thumb, index, middle). In Fig. 6.16a the user is unscrewing a cap from a bottle using only one hand. Ulnar grasp is used to keep firm the bottle, while the other fingers can unscrew the cap. Figure 6.16b shows the example of grasping big size box with the augmented hand that is impossible to grasp with the human hand only. Holding multiple objects with the augmented hand is shown in Fig. 6.16c, f. The example illustrated in Fig. 6.16d involves the task of opening the door using the handle, while carrying an heavy bag with the hand. The user was able to turn the handle to open the door using the robotic device, while keep holding the bag with the hand. The Fig. 6.16e is another example where the user can solder a circuit

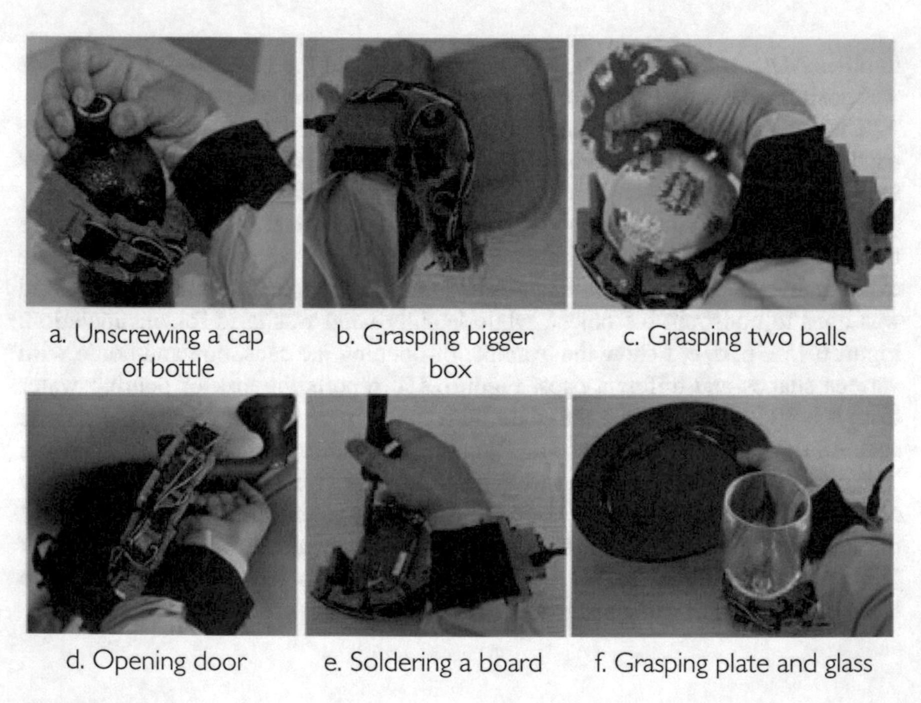


**Fig. 6.16** Examples of tasks performed by the augmented hand, i.e., human hand plus supernumerary robotic finger. In all the tasks, the human healthy hand and robotic finger work together to complete the tasks which are impossible to do with human hand only

board, while holding the board by robotic finger, ring and pinkie. The thumb, index and middle finger are used to hold soldering gun. Note that, all the tasks are either impossible or at least very difficult to be carry out with a single hand. All these tasks were successfully fulfilled by all the subjects with the help of the EMG interface and the supernumerary extra finger. Also in this subset of examples, the possibility to control both motion and joint stiffness of the device was exploited by the users.

## 6.5 Results and Discussion

In Sect. 6.2.1, we described the tasks performed by the subjects to prove the usability of the proposed EMG interface and the novel supernumerary finger prototype. In the following, we will give the details of the position of the device and the forces exerted on the grasped object for two particular type of grasps, i.e., power and precision grasps. Fig. 6.17a–d reported the behavior of the device during power and precision grasping, respectively. In particular, Fig. 6.17a, b refer to the power grasp reported in Fig. 6.15a, while Fig. 6.17c, d refer to the precision grasp reported in Fig. 6.16e. We report only these examples for the sake of brevity. Figure 6.17a–d represent the average of five repetitions of the same subject. To measure the forces exerted on the objects, we equipped each module of the extra finger with a Force Sensing Resistor (FSR) (408, Interlink Electronics Inc., USA). The user was asked to command the supernumerary finger till the grasp is obtained. Once the device was in contact with the object, the user increased the stiffness of the device by co-contracting his/her biceps, see Fig. 6.18a. The contraction of the biceps was read by the EMG interface and the value of $k_d$ was increased, see Fig. 6.18b. This variation produced a variation on the desired angle $q_{des}$ of the modules, while the read actual position of the modules remained the same due to the constrain imposed by the object, see Fig. 6.17b, d. The variation of the desired angles produces however an increase of the force exerted by the device onto the object, as shown in Fig. 6.17a, c. So that, by co-contracting the biceps the user can regulate the grasp tightness. As expected, in power grasps all the modules move of a similar angle so to wrap the object. All the modules also contribute to the grasp tightness applying force on the object. Differently, in precision grasp the fingertip module is the only module exerting force. The module motion is opposed to the direction of the other three modules so to leave the fingertip parallel to the hand.

The proposed EMG interface and the novel robotic extra finger prototype successfully enabled the users to complete all the targeted tasks both related to augmentation and compensation. The experiments proved that the presented system can be an effective aid both in augmenting the healthy human hand or in compensating its missing abilities in case of a disease. The proposed EMG control interface resulted to be intuitive and simple. The users were able to generate multiple control inputs without using sensorized gloves on human hand and were able to modulate the compliance of the robotic device in proportional to the EMG signal amplitude variations at biceps. Moreover, the upgraded version of the device with additional adduction/abduction

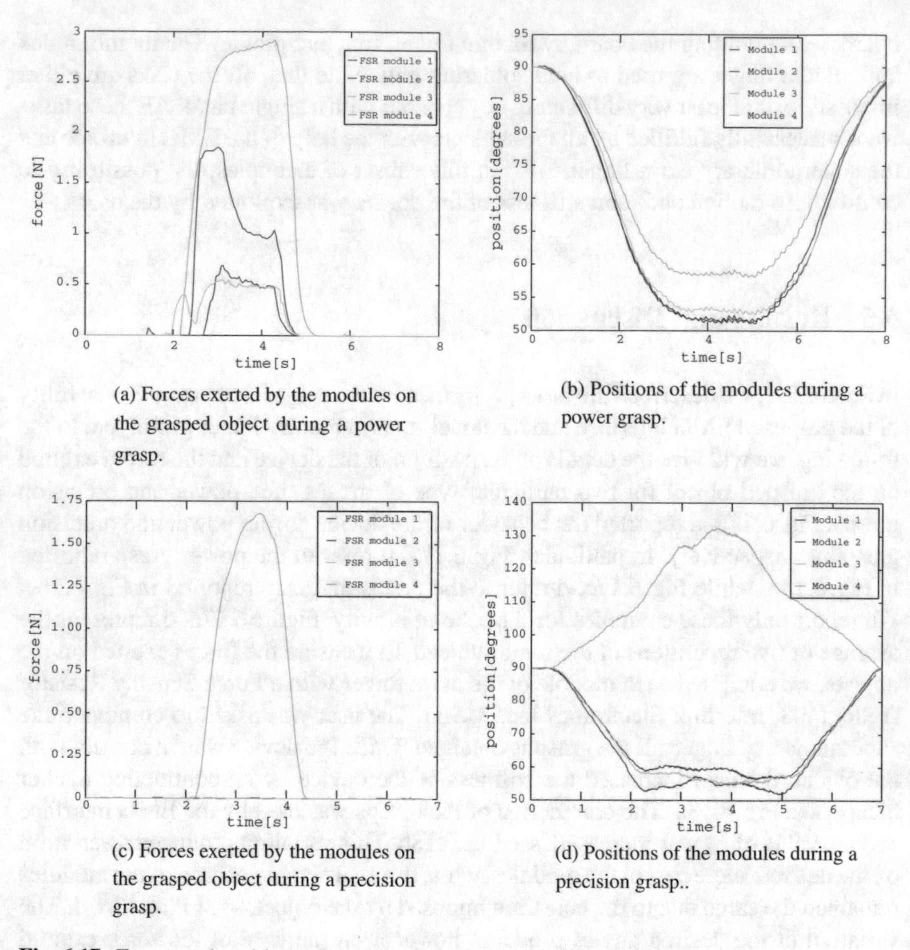


(a) Forces exerted by the modules on the grasped object during a power grasp.

(b) Positions of the modules during a power grasp.

(c) Forces exerted by the modules on the grasped object during a precision grasp.

(d) Positions of the modules during a precision grasp..

**Fig. 6.17** Forces and positions of modules in both power and precision grasp

degree of freedom increased the dexterity of the robotic device allowing more complex operation especially when hand augmentation was considered.

### *6.5.1 Conclusion*

The second part of this chapter presented an EMG control interface for a supernumerary robotic finger that can be used to control motion and joint stiffness. The aims are grasping compensation in chronic stroke patients and augmentation of human healthy hand to enhance its grasping capabilities and workspace. The motion of the robotic finger is controlled through gesture recognition and its compliance is regulated by EMG signal amplitude variations. In particular, we proposed Myo Armband

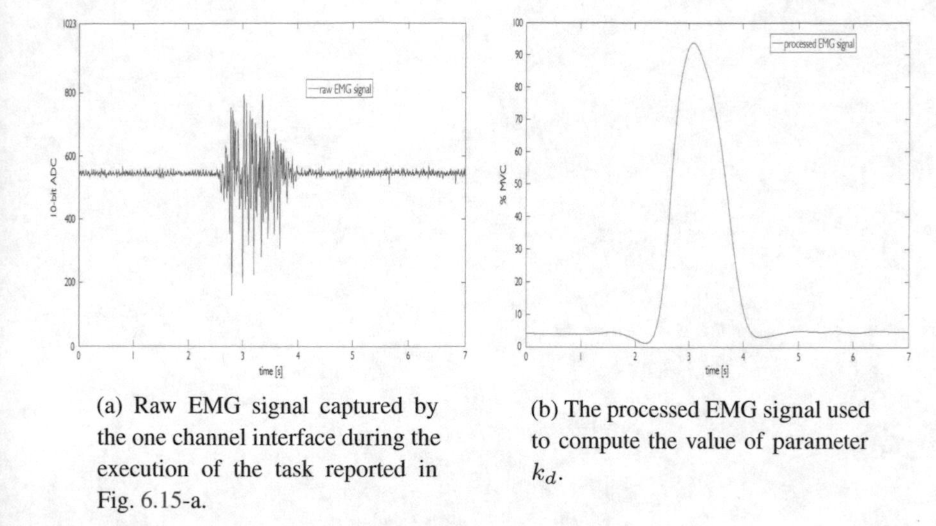


(a) Raw EMG signal captured by the one channel interface during the execution of the task reported in Fig. 6.15-a.

(b) The processed EMG signal used to compute the value of parameter $k_d$.

**Fig. 6.18** EMG signal for compliance regulation of robotic finger

to recognize the user gesture to control the motion of the robotic device. We developed EMG one channel electrode interface to modulate the compliance of the robotic device through a control scheme based on servo motor. We developed a five DoFs device that can be worn on the user wrist by an elastic band. We validated the use of device in augmenting and compensating the human hand grasping abilities. In particular, we showed how the supernumerary-robotic finger can play the role of an extra thumb enlarging the human hand workspace and the hand dexterity and how it can compensate the missing abilities of the non-functional hand in case of stroke patients. We demonstrate through experiments that the same interface can be used by patient and healthy subjects to control different flexion trajectories and to regulate the grasp tightness.

# Chapter 7
# From Grasp Compensation Towards Hemiparetic Upper Limb Rehabilitation

A high number of the chronic stroke patients do not recover a full mobility of their affected upper limb. Deficits in motor execution ranges from weakness of wrist/finger extensors to inefficient scaling of grip force and peak aperture. This means that both hand and arm impairments contribute to a low capability of grasping and manipulating objects. Patients with hemiparesis often have limited functionality in the paretic limb. The standard therapeutic approach requires the patient to attempt to make use of the paretic arm and hand even though they are not functionally capable, which can result in feelings of frustration and failure. Moreover, the hemiparetic patients also face challenges in completing many bimanual tasks in activity of daily living (ADL).

In this chapter, we present two robotic devices to provide the needed assistance during paretic upper limb rehabilitation involving both grasping and arm mobility to solve task-oriented activities. In particular, we propose the combination of the Soft-SixthFinger with the zero gravity arm support, the SaeboMAS. The Soft-SixthFinger is a wearable robotic supernumerary finger designed to be used as an active assistive device by post stroke patients to compensate the paretic hand grasp. The device works jointly with the paretic hand/arm to grasp an object similarly to the two parts of a robotic gripper. The SaeboMAS is mobile arm support to neutralize gravity effects on the paretic arm specifically designed to facilitate and challenge the weakened shoulder muscles during functional tasks. The proposed system has been designed to be used during the rehabilitation phase when the arm is potentially able to recover its functionality, but the hand is still not able to perform a grasp due to the lack of an efficient thumb opposition. The overall system also act as a motivation tool for the patients to perform task-oriented rehabilitation activities. With the aid of proposed system, the patient can closely simulate the desired motion with the non-functional arm for rehabilitation purposes, while performing a grasp with the help of the Soft-SixthFinger. As a pilot study we tested the proposed system with a chronic stroke patient to evaluate how the mobile arm support in conjunction with a robotic supernumerary finger can help in performing the tasks requiring the manipulation



I. Hussain and D. Prattichizzo, *Augmenting Human Manipulation Abilities with Supernumerary Robotic Limbs*, Biosystems & Biorobotics 26,
https://doi.org/10.1007/978-3-030-52002-1_7

of grasped object through the paretic arm. In particular, we performed the Frenchay Arm Test (FAT) and Box and Block Test (BBT). The proposed system successfully enabled the patient to complete tasks which were previously impossible to perform.

## 7.1 Introduction

Long-term disabilities of the upper limb affects millions of stroke survivors [27]. More than 80% of individuals who experience severe hemiparesis after stroke cannot completely recover hand and arm functionality [32]. The improvement of the paretic hand functionality plays a key role in the functional recovery of stroke patients with a paretic upper limb [70, 71]. Different motor impairments can affect the hand both at motor execution and motor planning/learning level ranging from weakness of wrist/finger extensors, increased wrist/finger flexors tone and spasticity, co-contraction, impaired finger independence, poor coordination between grip and load forces, inefficient scaling of grip force and peak aperture, and delayed preparation, initiation, and termination of object grip [30].

In the last two decades, several rehabilitation teams have started to integrate robotic-aided therapies in their rehabilitation projects. Such treatments represent a novel and promising approach in rehabilitation of the post-stroke paretic upper limb. The use of robotic devices in rehabilitation can provide high-intensity, repetitive, task-specific and interactive treatment of the impaired upper limb and can serve as an objective and reliable means of monitoring patient progress [74–76]. Most of the proposed devices for hand and arm rehabilitation are designed to increase functional recovery in the first period after the stroke when, in some cases, biological restoring and plastic reorganization of the central nervous system take place [31]. However, even after extensive therapeutic interventions in acute rehabilitation, the probability of regaining functional use of the impaired hand is low [155]. For this reason, we recently started investigating on robotic devices for the compensation of hand function in chronic stroke patients. In [23, 25, 36, 156] we introduced a wearable robotic extra finger that can be used as an active compensatory tool for grasping action by chronic stroke patients. The principle of use of the proposed extra finger is rather simple and intuitive. The device can be worn on the paretic forearm by means of an elastic band. The robotic finger and the paretic hand act like the two parts of a gripper working together to hold an object, see Fig. 7.1. This solution represents the minimum robotic complexity necessary to grasp and hold and object. The user is able to control the flexion/extension of the robotic finger through an EMG interface placed on the patient forehead [37]. In [2] we showed how the robotic sixth finger can be used in Activities of Daily Living (ADL) involving common bimanual tasks including opening cans and jars with different closing system and shapes.

In our preliminary experiments with patients, we noticed that compensation process by using extra finger motivates the patient to use her or his muscles to coordinate with the device for the completion of the task. Thus, the extra finger acts like an active and motivational assistance device. This approach encourages the patients

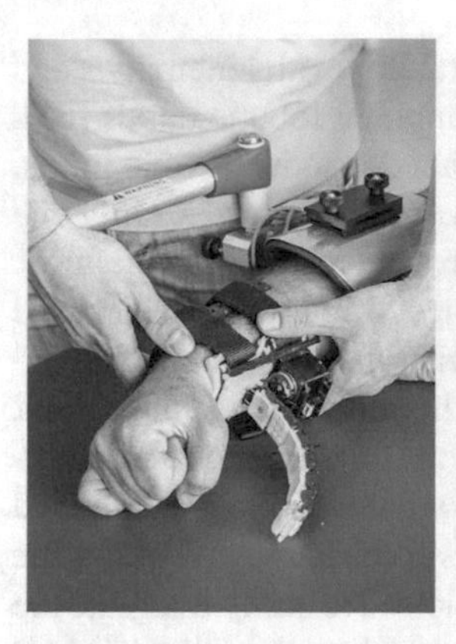

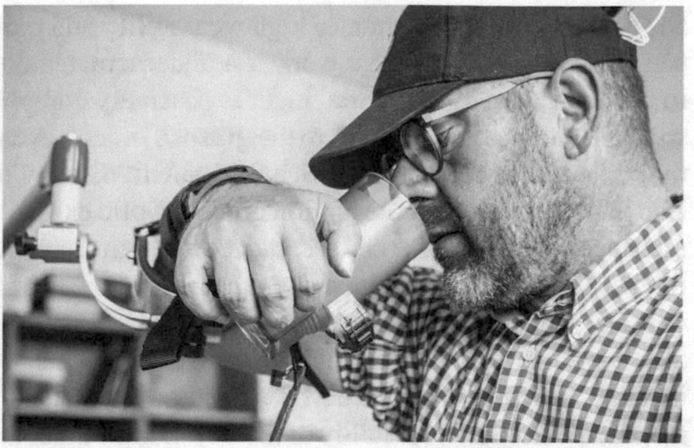

**Fig. 7.1** The proposed system: The Soft-SixthFinger works with the paretic upper limb to compensate for hand grasping functionality. The motion of the device is controlled by a wearable EMG interface embedded in a cap. The Passive arm support compensates the gravity force of the paretic upper limb

to use their potential and residual abilities effectively instead of being fully dependent on the motion of robotic device like passive assistive devices. The use of a robotic extra finger also limits the drawbacks of other compensatory strategies [78] that lead to a disuse of the affected arm and hand often lead to the learned non-use phenomenon of the hemiplegic upper extremity [151]. Based on these observations, we started exploring a possible use of the supernumerary finger as a augmentative device also during rehabilitation in acute and sub-acute phases. In particular, we consider a possible integration of the robotic sixth finger with a mobile arm support (MAS) to compensate for arm weight. Devices for compensating the gravity force of the arm can be used during stroke rehabilitation therapy to increase the quality and quantity of movements and reduce the fatigue made by patients with upper limb impairment [157–159], especially in the sub-acute post-stroke phases.

Supporting the weight of the arm is thought to benefit upper limb rehabilitation primarily by increasing capacity in terms of intensity or volume of therapeutic exercises [160]. In [161] a study on chronic stroke patients reported that gravity supported arm exercises can improve arm movement ability. A MAS counteracts the effects of gravity while facilitating and promoting functional movement. In addition to using a MAS for function, research has shown that gravity compensation devices are also effective for improving motor control, decreasing spasticity and minimizing fatigue [157–159].

In this work, we present a possible solution for an augmented motor rehabilitation which may increase the paretic upper limb recover of functions thanks to the combination of a robotic extra finger, called the Soft-SixthFinger [2], and the SaeboMAS (Saebo, Charlotte, USA). The aim of the Soft-SixthFinger is not to assist the paretic hand motion of the patient, but rather to add just what is needed to grasp: an extra thumb. The robotic extra finger is worn on the user's forearm and can accomplish a given task in cooperation with the paretic limb, see Fig. 7.1. The robotic extra finger has been designed to guarantee high wearability and portability with kinematics and actuation inspired by recent works on underactuated compliant robotic hands [65]. In particular, the robotic extra finger is passively compliant due to its flexible joints, so that it automatically adapts to the grasped object, even of different sizes. Only one motor is used to control the device flexion/extension with a tendon-driven actuation. The patient can control the motion of the robotic extra finger through an EMG based interface. Such interface can recognize, through the acquisition of the Electromyography (EMG) signal measured at the frontalis muscle of the patient, when the patient voluntary moves his or her eyebrows upwards. Frontalis muscle contractions generate events that regulate the finger flexion/extension. The whole system is embedded in a cap. Electrodes can be easily placed on the patient's forehead just wearing the interface. The high wearability of the Soft-SixthFinger allowed to easily integrate the device with the arm supports. The proposed system resemble the integration of MAS with active and passive orthoses [162, 163]. The clear advantages of our system is the ease of wear, the wearability and extreme portability due to a light weight and the ease of use. On the other side, hand orthoses also contribute to the rehabilitation of the hand and the wrist. Our system can be mainly used for rehabilitation of the arm as a way to promote arm motion even in case of weak proximal muscle when grasping function of the hand is not present yet.

To test the proposed system, we set up a pilot experiment where a chronic stroke patient was asked to perform two different tests: a modified box and block test [150] and the Frenchay Arm Test [81]. The purposes of the experiment were to verify the integration of the Soft-SixthFinger with the SaeboMAS and to validate with a patient possible exercises useful during rehabilitation.

The rest of the paper is organized as follows. In Sect. 7.2 all the parts composing the proposed system are described in details. Section 7.3 describes the performed experiments, while in Sect. 7.4 a discussion on the possible use of the proposed system is reported. Conclusion and future works are reported in Sect. 7.5.

## 7.2 The Proposed System

In this chapter we propose the Soft-SixthFinger to compensate the missing grasping abilities of paretic hand and the mobile arm support to help compensate for the effects of gravity of the impaired arm. Both devices are shown in Fig. 7.2 and the details of each device are presented in this section.

### 7.2.1 The Soft-SixthFinger and EMG Interface

The Soft-SixthFinger has been designed to compensate the missing grasping abilities of patients with neuromuscular disorder. The device is developed by robotic and rehabilitation teams taking into account the engineering design guidelines which

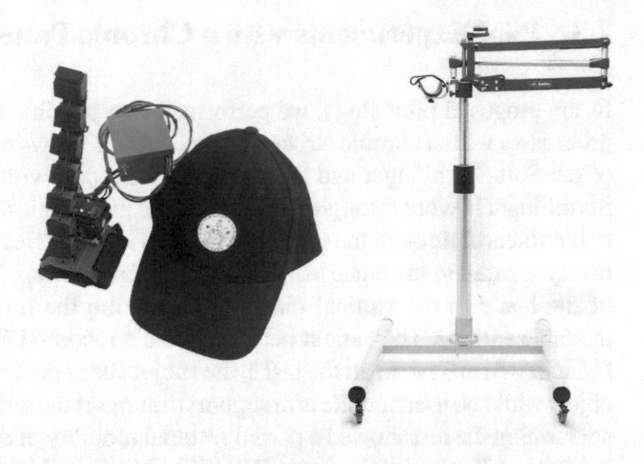

**Fig. 7.2** On left: Soft-SixthFinger, controller unit and eCap. On right: SaeboMAS for compensating the gravity of the arm

are suitable to clinical needs. In particular, wearability, modularity, lightweight and robustness are fundamental structural features of the device. The wearability concept applied to the design includes also ergonomics and compact size. In addition to that, passive compliance and flexibility for shape adaptation, ease of use, comfort and intuitive user control interface, are from the functional point of view other important features of the device. The detail design and development of the soft robotic finger [2, 164] are presented in Chap. 4. The motion of the supernumerary finger is controlled by EMG frontalis muscle interface detailed in Chap. 6. The patients can voluntary control the Soft-SixthFinger by contracting the frontalis muscle on their forehead. This muscle, due to a bilateral cortical representation, is always spared in case of a motor stroke, either of the left or of the right hemisphere. Activation of the muscle can be achieved by moving the eyebrows upwards. The bipolar EMG electrodes placement inside the cap provides twofold advantages, firstly the electrodes can easily be placed on the patient's frontalis muscle, secondly the stroke patients can easily wear the eCap using their healthy hand without requiring any external help.

### 7.2.2 *The Mobile Arm Support (SaeboMAS)*

The SaeboMAS is used for gravity compensation of the impaired upper limb, allowing the patient to perform exercises, as well as self-care activities demanding a smaller muscle effort. The device can be table-mounted moreover a height adjustable rolling base makes the SaeboMAS mobile for easy relocation throughout user facility. It has an elbow support and a comfortable malleable forearm support with removable liners for infection control. Through an adjustable spring based parallelogram the SaeboMAS can offer various levels of assistance to the user, while a measurable graded tension scale is useful for tracking and documenting progress.

## 7.3 Pilot Experiments with a Chronic Patient

In the proposed pilot study, we performed two possible tests to asses rehabilitation progresses with a chronic stroke patient. Fig. 7.1 shows on the left side the coupling of the Soft-SixthFinger and the mobile arm support with the paretic arm. The Soft-SixthFinger is worn through the support base and elastic straps. The MAS is attached to the forearm through the custom brace and its end-effector support. The forearm is firmly secured to the brace using elasticized fabric wrap. The MAS prevents rotation of the brace in the vertical plane thus ensuring the forearm is always parallel to the table surface. The patient performed the a modified Box and Block Test and the Frenchay Arm Test. In all the tasks, the subject used the Soft-SixthFinger to grasp the object while he used mobile arm support to support the weight of the arm. The subject performing the test showed a partial residual mobility of the arm ($\leq 2$ in the National Institute of Health Stroke Scale (NIHSS) [84], item 5 "paretic arm"). Moreover, the patient showed the following characteristics: normal consciousness (NIHSS, item 1a, 1b, 1c $= 0$), absence of conjugate eyes deviation (NIHSS, item $2 = 0$), absence of complete hemianopia (NIHSS, item $3 \leq 1$), absence of ataxia (NIHSS, item $7 = 0$), absence of completely sensory loss (NIHSS, item $8 \leq 1$), absence of aphasia (NIHSS, item $9 = 0$), absence of profound extinction and inattention (NIHSS, item $11 \leq 1$). The rehabilitation team assisted the subject during a training phase for each test. During this phase, the optimal position of the devices on the arm, according to the patient motor deficit, was evaluated. The subject did preliminary trails to see his comfort in using the robotic devices and to get familiar with the control interface.

### *7.3.1 The Box and Block Test*

The Box and Block test measures unilateral gross manual dexterity [165]. We slightly modified the test in order to let the patient use the proposed system. In particular, we removed the first compartment of classic setup leaving the cubes free on the table. The patient was then able to use the arm support plus the extra finger to grasp the cubes and move them inside the second compartment. The subject used the Soft-SixthFinger to grasp the blocks from the table while the arm support was used to assist the paretic arm mobility. The test setup is report in Fig. 7.3.

After the training phase, the patient performed the test three times with four different conditions: (a) without using any device, (b) only MAS, (c) only Soft-SixthFinger and (d) the complete system. The results of the test are reported in Table 7.1 and screenshots of the tasks are shown in Fig. 7.3. On average, patient scored 8 blocks per minute while using the complete system (condition d). On the contrary, he scored 0 while using none of the devices.

**Fig. 7.3** The box and block test: The subject grasped the object with the help of the soft-sixthfinger while using passive arm support to assist the paretic arm

**Table 7.1** Results of the Box and Block Test, without using any device, only MAS, only SSF and SSF plus MAS

| Box and Block Test | None | SSF only | MAS only | SSF plus MAS |
|---|---|---|---|---|
| Blocks per minute | 0 | 3 | 0 | 6 |

## 7.3.2 *The Frenchay Arm Test*

The test consists of five pass/fail tasks to be executed in less then three minutes. The patient scores 1 for each of the successfully completed task, while he or she scores 0 in case of fail. The subject sit at a table with his hands in his lap, and each task started from this position. He or she is then asked to use the affected arm/hand to:

1. *Task_1* Stabilize a ruler, while drawing a line with a pencil held in the other hand. To pass, the ruler must be held firmly.
2. *Task_2* Grasp a cylinder (12 mm diameter, 50 mm long), set on its side approximately 150 mm from the table edge, lift it about 300 mm and replace without dropping.
3. *Task_3* Pick up a glass, half full of water positioned about 150–300 mm from the edge of the table, drink some water and replace without spilling[1] (see Fig. 7.4).
4. *Task_4* Remove and replace a sprung clothes peg from a 10 mm diameter dowel, 150 mm long set in a 100 mm base, 150–300 mm from table edge. Not to drop peg or knock dowel over.
5. *Task_5* Comb hair (or imitate); must comb across top, down the back and down each side of head.

[1]Note that for safety reasons we did not use water in presence of electronic components.

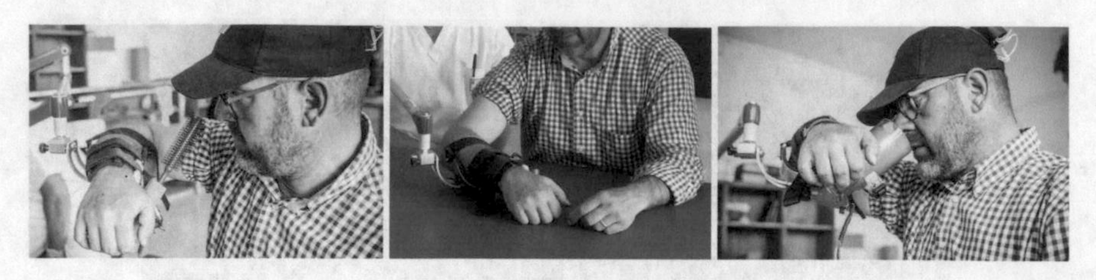

**Fig. 7.4** Three tasks of the Frenchay Arm Test fulfilled thanks to the proposed system: on the left comb hair, in the center grasp a cylinder and on the right pick up a glass

**Table 7.2** Results of the Frenchay Arm Test, without using any device, only MAS, only SSF and SSF plus MAS

| Frenchay Arm Test | None | SSF only | MAS only | SSF plus MAS |
|---|---|---|---|---|
| Stabilize a ruler | 1 | 1 | 1 | 1 |
| Grasp a cylinder | 0 | 1 | 0 | 1 |
| Pick up a glass | 0 | 1 | 0 | 1 |
| Remove a sprung | 0 | 0 | 0 | 0 |
| Comb hair | 0 | 0 | 0 | 1 |

When compared with other upper limb assessments, the Frenchay arm test has shown good reliability in measuring functional changes in stroke patients [81].

We performed the FAT with a chronic stroke patient using the proposed system to see how it can improve the performance of the subject. After the training phase, the subjects had three minutes to perform the test. The subjects performed the tasks with four different conditions, without using any device, only MAS, only SSF and SSF plus MAS. The order of tasks was selected randomly. The results of the test are reported in Table 7.2 and screenshots of the tasks are shown in Fig. 7.4. Patient scored 3 out of 3 while using complete system, on contrary he scored only 0 without using the proposed devices.

## 7.4 Discussion

Stroke survivors have to reacquire a very high level of hand motor control before they actually can use the limb in ADL. This might explain why stroke patients that seems to show an adequate movement ability when monitored in the laboratory often do not use the limb into ADL with the expected regularity [166, 167]. With the aid of the robotic finger and of a passive arm support, the patients can better perform rehabilitation tasks and closely simulate the desired motion with the non-functional arm for rehabilitation purposes. The passive arm support can assist patient's paretic arm in making more meaningful movements during the rehabilitation process, even when proximal arm muscles are still deficitary. It counteracts the effects of gravity to functionally integrate the patient's affected arm during various tasks. The patient can exercise the impaired arm by positioning the extra robotic finger next to the object and can attempt to use their paretic arm to complete the tasks to the extent of their maximum abilities with minimum assistance from the robotic devices. The level of assistance to paretic arm can be adjusted by setting the stiffness of the MAS after monitoring the patient improvement. The tasks can be completed by the robotic finger and with the help of arm support even if the patient's upper limb is too weak to do so, thus providing the patient with the motivation to attempt these tasks and minimizing the feelings of defeat that may arise when attempting to complete tasks beyond their current capability. This could help the patient to eventually regain the ability to complete these tasks on their own. The patient may benefit in the short term with increased feelings of independence while simultaneously working toward the long term goals of rehabilitation and healing.

The proposed system aims to assist persons who are not able to make a functional grasp with their paretic arm and hand themselves. The creation of a functional grasp by means of the extra-finger flexion enables patients to execute task-oriented grasp and release exercises and practice intensively using repetitive movements. Repetition is an important principle in motor learning which reflects the Hebbian learning rule that connections between neurons are strengthened when they are simultaneously active (i.e., long term potentiation) [168]. Repetitive task training is a key modality of effective training in stroke [169]. The proposed dynamic system can be used in goal-directed activities and lowers the threshold for patients to participate in a greater variety of evidence-based treatment programs. In fact, notwithstanding the proven efficacy of classic CIMT [86], the obvious impossibility to perform bimanual tasks in daily living activities may deeply affect patients' motivation. This issue is overcome by our device, which instead encourages and motivates the patients to fully use their residual abilities, meanwhile possibly acquiring new grasping abilities. For these reasons, we could even say that the MAS plus Soft-SixthFinger device touches in parallel both cognitive and motoric aspects of the rehabilitation process. From the perspective of therapy efficiency and patient satisfaction, the use of an extra finger may be beneficial, because less individual therapy assistance is needed. To date, most clinical studies reporting the use of compensatory tools have been performed in chronic stroke patients. A clinical assumption is that the benefits of such an extra finger may be larger in sub-acute stroke patients, because learned non-use and secondary complications like contractures may be prevented. However, this has not yet been investigated systematically. Obviously, stroke is not the only field of application of the proposed device. All diseases leading to upper limb/arm paresis (myelopathies, amyotrophic lateral sclerosis, multiple sclerosis, muscular dystrophies) may be theoretically a suitable ground of application.

The patients that could take advantage by using our proposed system may be those that show mild upper limb paresis associated with higher hand and fingers weakness in the first weeks after stroke, i.e., during the recovery phase [170]. Although the upper limb paresis is not complete, some of these patients are not able to move the arm against gravity, especially in the sagittal plane and, in addition, the absence of the grasping function makes impossible to perform a real "task-oriented therapy". For example, Nijland et al. [171] showed that at nine days after stroke some patients showed the shoulder abduction movement, but the fingers extension, which is crucial for the grasping function, is absent. The presence of the shoulder abduction movement is not usually highly correlated with the ability to move the arm against gravity in the sagittal plane in which the most of the ADL take place. In this case study, we have used the MAS mainly for helping the patient, who was able to perform the shoulder abduction (frontal plane), to better perform shoulder flexion against gravity (sagittal plane) and to grasp objects for solving tasks with the Soft-SixthFinger due to the absence of hand and fingers movements.

## 7.5 Conclusion

In this chapter, we presented an approach to use grasp compensatory device and passive arm support in rehabilitation exercise for hemiparetic upper limb where grasping an object and its manipulation through arm is fundamental. We propose the use of our combined system for an "augmenting rehabilitation approach" in addition to the common physical therapy in those patients in which some upper limb movements are present, but an "ecological" and task-oriented therapy cannot be performed. Indeed, there are evidence that better outcome is favored by intensive high repetitive task-oriented and task-specific training during the recovery phase [172]. We presented our preliminary results in combining the Soft-SixthFinger with the zero gravity arm support, SaeboMAS. In our previous works on active tools for manipulation compensation we focused mostly on the grasping part developing an extra-finger that can adapt to different object shapes. However, we noticed that most of the patients testing our devices were still not able to perform basic rehabilitation tests designed for upper limb due to the poor mobility of the paretic arm/shoulder. In this chapter we tested the feasibility of the proposed system with one of our same chronic stroke patients. As a result the patient was able to perform the tasks which even needed the grasped object manipulation through paretic arm mobility. The proposed system is a first step toward the realization of an assisting platform when the arm is recovering its functionality, but the hand is still not able to perform a grasp.

# Chapter 8
# Conclusions and Future Work

This thesis presents my contribution to the foundation of a new era of robotic extra-limbs, collecting all the work I have done toward my Ph.D. degree at the University of Siena. The focus of this research was on augmenting and compensating the human manipulation abilities through supernumerary robotic fingers. During these years, I investigated through various studies involving the robotics, biomechanics, human factors and control theory methods to develop new generation of supernumerary robotic fingers along with their customized control interfaces. The guiding principles of the robotic devices design are safety, wearability, ergonomics and user comfort.

The thesis presented in Chap. 1 an introduction to the supernumerary robotic fingers and state of the art. It illustrated how the research on the supernumerary robotic limbs is different in nature than the existing approaches of assistive devices. This was followed by a literature review on extra robotic fingers. It was understood that it's a newly introduced field which is extending its horizon from compensation to augmentation and enhancement of human manipulation abilities. The challenges lie in designing these robotic devices to fullfill the ergonomics and functional requirements as well as their integration with human body.

Chapter 2 has presented the design of extra robotic fingers to enhance the abilities of human hand to increase its work space and manipulation abilities. It proposed the different design guidelines both for fullyactuated and underactuated robotic extra-fingers that can be integrated with human hand to enhance its manipulation abilities. Such guidelines were followed for the realization of three prototypes obtained using rapid prototyping techniques, i.e., a 3D printer and an open hardware development platform. Both fully actuated and underactuated solutions have been explored. To control the motion of the robotic fingers, we presented mapping algorithm able to transfer to the extra-fingers a part or the whole motion of the human hand. The mapping algorithm was based on the definition of a virtual object obtained as a function of a set of reference points placed on the augmented hand (human hand and robotic fingers). When the human hand fingers were moved, the virtual object




I. Hussain and D. Prattichizzo, *Augmenting Human Manipulation Abilities with Supernumerary Robotic Limbs*, Biosystems & Biorobotics 26,
https://doi.org/10.1007/978-3-030-52002-1_8

was moved and deformed. The robotic extra-fingers were then actuated so that the reference points on them followed the same transformation. Although, The mapping algorithm allowed to move the extra-fingers according to the human hand motions without requiring explicit command by the user but it has some limitations too. This mapping approach required an instrumented glove to track the human hand which affect its practical application specially with stroke patients. Patients with a paretic hand cannot properly control finger motions, thus a dataglove interface cannot be used. The estimation of the human hand posture and fingers motion implies a reliable and computationally expensive hand tracking. Moreover, there is no feedback of the amount of the forces applied by the robotic finger on the grasped object and its status.

To overcome these limitation, in Chap. 5 the wearable sensory motor interfaces for supernumerary robotic fingers are presented. In particular, two kinds of interfaces namely "vibrotactile ring" and hRing are proposed. The human user is able to control the motion of the robotic finger through trigger signals generated by a switch placed on the rings, while being provided with cutaneous feedback about the forces exerted by the robotic finger on the environment. A trigger based control approach can dramatically simplifies the interaction with the extra robotic finger, reducing it to the activation of a grasping procedure through the trigger signal activated by a wearable switch placed on a ring. A single switch activation regulated the stop/motion of the finger along a predefined flexion trajectory, while a double activation switched from flexion to extension and viceversa. Although, the ring based control approach resulted simple and intuitive, this control interface involved human hand thumb. Thus limiting the use of thumb in completion of tasks. Moreover, it offers few user control inputs to control the motion of the robotic finger and force control is not straightforward.

Chapter 3 presents how the superumerary robbotic finger can be used to compensate the missing abilities on impaired upper limb in chronic stroke patients. The robotic finger and paretic limb act as two parts of a gripper to grasp and hold the object. With the use of supernumerary robotic fingers, It's expected to increase patients' performances, with a focus on objects manipulation, thereby improving their independence in ADL, and simultaneously decreasing erroneous compensatory motor strategies for solving everyday tasks. The patient was control the motion of the robotic finger using an EMG interface through frantalis muscles.

The use of preliminary prototype of the robotic finger helped us in understanding the potential of the approach in case of stroke patients and extracting the ergonomics and functional requirements. The fully actuated structure allows the robotic finger to actively shape around an object, but the resulting size and weight affect the device portability, wearability and robustness. For these reasons, we proposed soft and underactuated robotic fingers. In particular, Chap. 4 presents the design, analysis, fabrication, experimental characterization and evaluation of two prototypes of soft supernumerary robotic fingers that can be used as grasp compensatory devices for hemiparetic upper limb. We proposed a method to compute the stiffness of flexible joints and its realization in order to let the fingers track a certain predefined trajectory. We referred to tendon driven, underactuated and passively compliant fingers composed of deformable joints and rigid links. The finger joints can be given specific stiffness and pre-form shapes such that a single cable actuation can be used. We define firstly a procedure to determine suitable joints stiffness and then we propose a possible realization in robotics fingers hardware structure. The stiffness computation is obtained leveraging on the the mechanics of tendon-driven hands and on compliant systems, while for its implementation beam theory has been exploited. We validate the proposed framework both in simulation and with experiments using a prototypes of the devices.

In Chap. 6 two kinds of electromyographic (EMG) control interfaces for supernumerary robotic fingers are presented. In particular, frontalis muscle cap and arm EMG interface. The former is more suitable for underactuated soft fingers while the later is developed to control the motion and compliance of fully actuated robotic finger. In frontalis muscle cap the electrodes and acquisition boards are embedded in cap which allows the user to control the device motion through wireless communication by contracting the frontalis muscle.

Finally, Chap. 7 presents the combination of soft extra finger with the arm support to provide the needed assistance during paretic upper limb rehabilitation involving both grasping and arm mobility to solve task-oriented activities. The proposed system has been designed to be used during the rehabilitation phase when the arm is potentially able to recover its functionality, but the hand is still not able to perform a grasp due to the lack of an efficient thumb opposition. The overall system also act as a motivation tool for the patients to perform task-oriented rehabilitation activities.

Currently, we are investigating whether it is possible to introduce the robotic extra fingers in early rehabilitation phase. Many of the rehabilitation strategies involve ADL tasks where patients attempt to make use of weak hand even though it is not functionally capable. This can result in feelings of frustration. Presenting an active compensatory tool may help in the initial phase to promote the use of the arm even if the hand grasp function is not recovered. Furthermore, hemiparetic patients also face challenges in completing many bimanual tasks which are crucial to their independence and quality of life , some of examples are shown in this book. Additional opportunities for these robotic devices include patient monitored rehabilitation training that can be done at home and direct ADL assistance for those with minimal hand functionality. For example, some studies in literature claim that rehabilitation in chronic stages should focus on learning adaptive processes either through more difficult bi-manual activities or through the forced use of the affected limb [173]. In this regard, the patients can use their muscles abilities effectively to complete the bi-manual task, while exploiting only minimal assistance by the robotic devices needed to complete the tasks. Moreover, clinical team could develop new protocols and scenarios for the involvement of such robotic compensatory devices, while monitoring the patients improvements on regular basis. The patient could exercise the impaired arm by positioning the grasp compensatory device next to the object, trying to contribute in grasping to the maximum extent of their abilities. This would motivate the patient to attempt the bi-manual tasks and minimize the feelings of failure that may arise when attempting to complete tasks beyond their current capability. This could help the patient to eventually regain the ability to complete these tasks on their own. The patient may benefit in the short term with the increased feelings of independence, while simultaneously working toward the long term goals of rehabilitation.

In future, we will also explore the possible use of the supernumerary robotic fingers during the rehabilitation phase, possibly integrating them with other state-of-the-art assistive devices used for hand rehabilitation, e.g. orthoses and functional electrical stimulation (FES). In a possible hand rehabilitation scenario, an object could be connected to the tip of the finger and actively moved with respect to the palm of the patients impaired hand. Thus, while the user is trying to stabilize the grasp, the robotic finger will have the aim of destabilizing it. Challenging grasping exercises like this, will stimulate movements of the impaired hand, possibly achieving a rehabilitative effect.

We are also exploring the possibility of using our devices in patients affected by other neurological diseases possibly affecting hand grasping, such as Multiple Sclerosis, Amyotrophic Lateral Sclerosis and paresis due to cervical spinal cord lesions.

We are also working on neural embodiment of supernumerary robotic fingers but this study is still at preliminary stage. The magnetic resonance imaging (MRI) compatible finger is needed to perform these neuroscience studies of robotic extra fingers [174]. Functional brain exploration methodologies, such as functional Magnetic Resonance Imaging (fMRI), are at present used to study perceptual and cognitive processes. To develop more complex experimental fMRI paradigm, researchers are interested in realizing active interfaces, using electrically powered actuators and sensors to be used inside the MRI environment [175].The use of non-ferromagnetic metals with higher stiffness and rigidity compared to plastic facilitates the design of smaller devices. In literature, several reports provide criteria for MR compatible devices.The experimental protocols are described to evaluate compatibility,in addition, it defines location and timing zones, where MR compatibility with respect to each zone should be stated. Also the various phenomena that can happen when a mechatronic device is placed adjacent to MRI scanner and is driven it during imaging.

MRI compatibility needs a set of supplementary constraints in the actuator and sensor choice and requires a proper design process. The common standard mechanical parts cannot be used in magnetic resonance environment because they usually contain ferromagnetic components. Majority of the actuators are electromagnetic devices and practically impossible to use during imaging.

We developed MRI compatible robotic fingers for preliminary experiments inside FMRI by properly choosing materials like ABS, whose properties are compliant with this specific application and actuators like piezoelectric motors with properly shielded wires.

Currently, we are in process of performing fMRI experiments using MRI compatible finger for the study of neural embodiment of extra finger. The Fig. 8.1a shows the experiment setup during the use of robotic finger inside the scanner. The under-actuatde robotic finger (see, Fig. 8.1b) is MRI compatible and realized through ABS material and using piezoelectric motor. Figure 8.1c, d show examples of fMRI brain scan images, while the method used is detailed in Table 8.1.

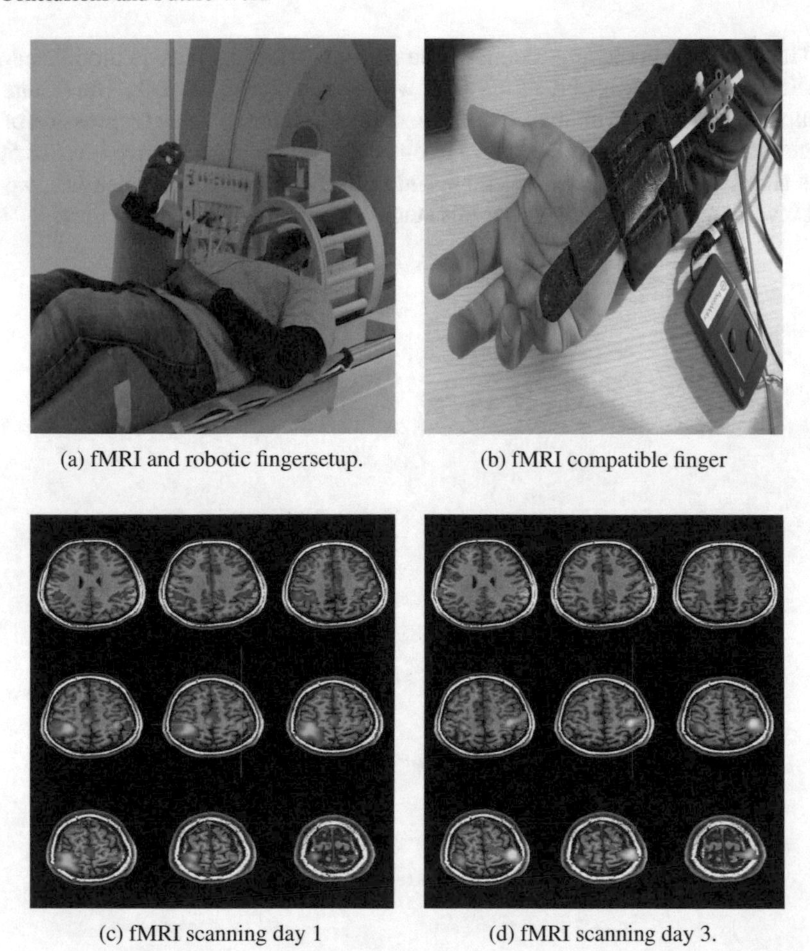

(a) fMRI and robotic fingersetup.

(b) fMRI compatible finger

(c) fMRI scanning day 1

(d) fMRI scanning day 3.

**Fig. 8.1** The fMRI experiments with MRI compatible finger having ABS material and piezoelectric motor

**Table 8.1** Methods

| 1.5 T siemens symphony scanner |
|---|
| EPI images, 4 runs (2 left hand, 2 right hand) |
| Finger tapping task (20 s for each finger; 4 blocks per run) |
| Analysis: standard preprocessing (AFNI) |
| Separate GLMs for the three sessions to obtain brain activation related to all the fingers |
| Alignment of the anatomical images to the one acquired during session 2 (FSL, rigid-body) |
| Warping of the functional results on the session 2 anatomical image |

The encoding procedure identified the regions whose activity is modulated and predictable on the basis of the finger that was moved during the task. The results are a map of the most discriminative voxels and a final accuracy value (expression of the success of the procedure). The leave-two-out procedure has a chance level of 50%. This study is still at preliminary stage and with the help of neuroscientists, we are still investigating the protocols for this study and analyzing the initial data.

# References


1. Bonilla BL, Asada HH (2014) A robot on the shoulder: coordinated human-wearable robot control using coloured petri nets and partial least squares predictions. In: 2014 IEEE international conference on robotics and automation (ICRA), pp 119–125
2. Hussain I, Salvietti G, Spagnoletti G, Prattichizzo D (2016) The soft-sixthfinger: a wearable emg controlled robotic extra-finger for grasp compensation in chronic stroke patients. IEEE Robot Autom Lett 1(2):1000–1006
3. Toledo C, Leija L, Munoz R, Vera A, Ramirez A (2009) Upper limb prostheses for amputations above elbow: a review. In: 2009 Pan American health care exchanges. IEEE, pp 104–108
4. Nowak DA (2008) The impact of stroke on the performance of grasping: usefulness of kinetic and kinematic motion analysis. Neurosci Biobehav Rev 32(8):1439–1450
5. Bogue R (2009) Exoskeletons and robotic prosthetics: a review of recent developments. Ind Robot: Int J 36(5):421–427
6. Stefanov DH, Bien Z, Bang W-C (2004) The smart house for older persons and persons with physical disabilities: structure, technology arrangements, and perspectives. IEEE Trans Neural Syst Rehabil Eng 12(2):228–250
7. Van der Loos HFM, Reinkensmeyer DJ (2008) Rehabilitation and health care robotics. Springer handbook of robotics. Springer, Berlin, pp 1223–1251
8. Mohammed S, Amirat Y (2009) Towards intelligent lower limb wearable robots: challenges and perspectives - state of the art. In: IEEE international conference on robotics and biomimetics, 2008. ROBIO 2008, pp 312–317
9. Pons JL, et al (2008) Wearable robots: biomechatronic exoskeletons, vol. 338. Wiley Online Library
10. Masia L, Hussain I, Xiloyannis M, Pacchierotti C, Cappello L, Malvezzi M, Spagnoletti G, Antuvan C, Khanh D, Pozzi M, et al (2018) Soft wearable assistive robotics: exosuits and supernumerary limbs
11. Belter JT, Segil JL (2013) Mechanical design and performance specifications of anthropomorphic prosthetic hands: a review. J Rehabil Res Dev 50(5):599
12. Brokaw EB, Black I, Holley RJ, Lum PS (2011) Hand spring operated movement enhancer (handsome): a portable, passive hand exoskeleton for stroke rehabilitation. IEEE Trans Neural Syst Rehabil Eng 19(4):391–399




I. Hussain and D. Prattichizzo, *Augmenting Human Manipulation Abilities with Supernumerary Robotic Limbs*, Biosystems & Biorobotics 26,
https://doi.org/10.1007/978-3-030-52002-1

13. Davenport C, Parietti F, Asada HH (2012) Design and biomechanical analysis of supernumerary robotic limbs. In: ASME 2012 5th annual dynamic systems and control conference joint with the JSME 2012 11th motion and vibration conference, pp 787–793
14. Parietti F, Asada H (2016) Supernumerary robotic limbs for human body support. IEEE Trans Rob 32(2):301–311
15. Parietti F, Chan KC, Hunter B, Asada HH (2015) Design and control of supernumerary robotic limbs for balance augmentation. In: 2015 IEEE international conference on robotics and automation (ICRA). IEEE, pp 5010–5017
16. Elahe A, Etienne B, Mohamed B, Hannes B (2015) Control of a supernumerary robotic hand by foot: an experimental study in virtual reality. PloS One 10(7):
17. Treers L, Lo R, Cheung M, Guy A, Guggenheim J, Parietti F, Asada H (2016) Design and control of lightweight supernumerary robotic limbs for sitting/standing assistance. In: International symposium on experimental robotics. Springer, pp 299–308
18. Khodambashi R, Weinberg G, Singhose W, Rishmawi S, Murali V, Kim E (2016) User oriented assessment of vibration suppression by command shaping in a supernumerary wearable robotic arm. In: 2016 IEEE-RAS 16th international conference on humanoid robots (Humanoids). IEEE, pp 1067–1072
19. Parietti F, Asada HH (2013) Dynamic analysis and state estimation for wearable robotic limbs subject to human-induced disturbances. In: IEEE international conference on robotics and automation (ICRA). IEEE, pp 3880–3887
20. Kurek DA, Asada HH (2017) The mantisbot: design and impedance control of supernumerary robotic limbs for near-ground work. In: 2017 IEEE international conference on robotics and automation (ICRA). IEEE, pp 5942–5947
21. Llorens-Bonilla B, Parietti F, Asada HH (2012) Demonstration-based control of supernumerary robotic limbs. In: IEEE/RSJ international conference on intelligent robots and systems (IROS). IEEE, pp 3936–3942
22. Wu F, Asada H (2014) Bio-artificial synergies for grasp posture control of supernumerary robotic fingers. In: Proceedings of robotics: science and systems, Berkeley, USA
23. Prattichizzo D, Salvietti G, Chinello F, Malvezzi M (2014) An object-based mapping algorithm to control wearable robotic extra-fingers. In Proceedings of IEEE/ASME international conference on advanced intelligent mechatronics, Besançon, France
24. Wu FY, Asada HH (2016) Implicit and intuitive grasp posture control for wearable robotic fingers: a data-driven method using partial least squares. IEEE Trans Rob 32(1):176–186
25. Prattichizzo D, Malvezzi M, Hussain I, Salvietti G (2014) The sixth-finger: a modular extra-finger to enhance human hand capabilities. In Proceedings of the IEEE international symposium on robot and human interactive communication, Edinburgh, United Kingdom
26. Hussain I, Meli L, Pacchierotti C, Salvietti G, Prattichizzo D (2015) Vibrotactile haptic feedback for intuitive control of robotic extra fingers. World haptics, pp 394–399
27. Go AS, Mozaffarian D, Roger VL, Benjamin EJ, Berry JD, Blaha MJ, Dai S, Ford ES, Fox CS, Franco S et al (2014) Heart disease and stroke statistics-2014 update: a report from the american heart association. Circulation 129(3):e28
28. Nowak DA (2008) The impact of stroke on the performance of grasping: usefulness of kinetic and kinematic motion analysis. Neurosci Biobehav Rev 32(8):1439–1450
29. Raghavan P, Krakauer JW, Gordon AM (2006) Impaired anticipatory control of fingertip forces in patients with a pure motor or sensorimotor lacunar syndrome. Brain 129(6):1415–1425
30. Balasubramanian S, Klein J, Burdet E (2010) Robot-assisted rehabilitation of hand function. Curr Opin Neurol 23(6):661–670
31. Lum PS, Godfrey SB, Brokaw EB, Holley RJ, Nichols D (2012) Robotic approaches for rehabilitation of hand function after stroke. Am J Phys Med Rehabil 91(11):S242–S254
32. Nakayama H, Jorgensen HS, Raaschou HO, Olsen TS (1994) Compensation in recovery of upper extremity function after stroke: the copenhagen stroke study. Arch Phys Med Rehabil 75(8):852–857
33. Michaelsen SM, Jacobs S, Roby-Brami A, Levin MF (2004) Compensation for distal impairments of grasping in adults with hemiparesis. Exp Brain Res 157(2):162–173

34. Aszmann OC, Roche AD, Salminger S, Paternostro-Sluga T, Herceg M, Sturma A, Hofer C, Farina D (2015) Bionic reconstruction to restore hand function after brachial plexus injury: a case series of three patients. The Lancet 385(9983):2183–2189
35. Pons JL, Rocon E, Ceres R, Reynaerts D, Saro B, Levin S, Van Moorleghem W (2004) The manus-hand dextrous robotics upper limb prosthesis: mechanical and manipulation aspects. Auton Robot 16(2):143–163
36. Hussain I, Salvietti G, Meli L, Pacchierotti C, Cioncoloni D, Rossi S, Prattichizzo D (2015) Using the robotic sixth finger and vibrotactile feedback for grasp compensation in chronic stroke patients. In: 2015 IEEE international conference on rehabilitation robotics (ICORR). IEEE, pp 67–72
37. Salvietti G, Hussain I, Cioncoloni D, Taddei S, Rossi S, Prattichizzo D (2016) Compensating hand function in chronic stroke patients through the robotic sixth finger. IEEE Trans Neural Syst Rehabil Eng 25(2):142–150
38. Odhner LU, Jentoft LP, Claffee MR, Corson N, Tenzer Y, Ma RR, Buehler M, Kohout R, Howe RD, Dollar AM (2014) A compliant, underactuated hand for robust manipulation. Int J Robot Res 33(5):736–752
39. Grebenstein M, Chalon M, Friedl W, Haddadin S, Wimböck T, Hirzinger G, Siegwart R (2012) The hand of the dlr hand arm system: designed for interaction. Int J Robot Res 31(13):1531–1555
40. Botvinick M, Cohen J et al (1998) Rubber hands' feel'touch that eyes see. Nature 391(6669):756–756
41. Guterstam A, Petkova VI, Ehrsson HH (2011) The illusion of owning a third arm. PloS One 6(2):e17208
42. van der Hoort B, Guterstam A, Ehrsson HH (2011) Being barbie: the size of one's own body determines the perceived size of the world. PloS One 6(5):e20195
43. Newport R, Pearce R, Preston C (2010) Fake hands in action: embodiment and control of supernumerary limbs. Exp Brain Res 204(3):385–395
44. Folegatti A, Farne A, Salemme R, De Vignemont F (2012) The rubber hand illusion: two'sa company, but three'sa crowd. Conscious Cogn 21(2):799–812
45. Higuchi S, Chaminade T, Imamizu H, Kawato M (2009) Shared neural correlates for language and tool use in broca's area. NeuroReport 20(15):1376–1381
46. Abdi E, Burdet E, Bouri M, Bleuler H (2015) Control of a supernumerary robotic hand by foot: an experimental study in virtual reality. PLoS One 10(7):e0134501
47. Zoss AB, Kazerooni H, Chu A (2006) Biomechanical design of the berkeley lower extremity exoskeleton (bleex). IEEE/ASME Trans Mechatron 11(2):128–138
48. Gupta A, O'Malley MK (2006) Design of a haptic arm exoskeleton for training and rehabilitation. IEEE/ASME Trans Mechatron 11(3):280–289
49. Atassi OA (2011) Design of a robotic finger for grasping enhancement. Master's thesis, Università degli Studi di Siena
50. Hussain I, Salvietti G, Malvezzi M, Prattichizzo D (2015) Design guidelines for a wearable robotic extra-finger. In: 2015 IEEE 1st international forum on research and technologies for society and industry leveraging a better tomorrow (RTSI), pp 54–60
51. Kiguchi K, Tanaka T, Fukuda T (2004) Neuro-fuzzy control of a robotic exoskeleton with emg signals. IEEE Trans Fuzzy Syst 12(4):481–490
52. Gioioso G, Salvietti G, Malvezzi M, Prattichizzo D (2013) Mapping synergies from human to robotic hands with dissimilar kinematics: an approach in the object domain. IEEE Trans Robot
53. Wu G, Van der Helm FCT, Veeger HEJ, Makhsous M, Van Roy P, Anglin C, Nagels J, Karduna AR, McQuade K, Wang X (2005) Isb recommendation on definitions of joint coordinate systems of various joints for the reporting of human joint motion-part ii: shoulder, elbow, wrist and hand. J Biomech 38(5):981–992
54. Gioioso G, Salvietti G, Malvezzi M, Prattichizzo D (2012) An object-based approach to map human hand synergies onto robotic hands with dissimilar kinematics. Robotics science and systems VIII. The MIT Press, Sydney

55. Salvietti G, Malvezzi M, Gioioso G, Prattichizzo D (2014) On the use of homogeneous transformations to map human hand movements onto robotic hands. In: Proceedings of IEEE international conference on robotics and automation, number 0, Hong Kong, China
56. Bicchi A (1995) On the closure properties of robotic grasping. Int J Robot Res 14(4):319–334
57. Murray RM, Li Z, Sastry SS (1994) A mathematical introduction to robotic manipulation
58. Gabiccini M, Bicchi A, Prattichizzo D, Malvezzi M. On the role of hand synergies in the optimal choice of grasping forces. Auton Robot, pp 1–18
59. Prattichizzo D, Malvezzi M, Gabiccini M, Bicchi A (2013) On motion and force controllability of precision grasps with hands actuated by soft synergies. IEEE Trans Robot in press:1–17
60. Kim B-H, Oh S-R, Yi B-J, Suh IH (2001) Optimal grasping based on non-dimensionalized performance indices. In: 2001 IEEE/RSJ international conference on intelligent robots and systems, 2001. Proceedings, vol 2. IEEE, pp 949–956
61. Buchholz B, Thomas JA, Streven AG (1992) Anthropometric data for describing the kinematics of the human hand. Ergonomics 35(3):261–273
62. Jones LA, Lederman SJ (2006) Human hand function. Oxford University Press, Oxford
63. Sanfilippo F, Salvietti G, Zhang H, Hildre HP, Prattichizzo D (2012) Efficient modular grasping: an iterative approach. In: Proceedings of IEEE international conference on biomedical robotics and biomechatronics. Rome, Italy, pp 1281–1286
64. Bicchi A, Prattichizzo D (2000) Manipulability of cooperating robots with unactuated joints and closed-chain mechanisms. IEEE Trans Robot Autom 16(4):336–345 August
65. Dollar AM, Howe RD (2011) Joint coupling design of underactuated hands for unstructured environments. Int J Robot Res 30(9):1157–1169
66. Birglen L, Gosselin CM (2004) Kinetostatic analysis of underactuated fingers. IEEE Trans Robot Autom 20(2):211–221
67. Immersion Technologies. Cyberglove wireless system. http://www.cyberglovesystems.com/
68. Sunderland A, Fletcher D, Bradley L, Tinson D, Hewer RL, Wade DT (1994) Enhanced physical therapy for arm function after stroke: a one year follow up study. J Neurol, Neurosurg Psychiatry 57(7):856–858
69. Wade DT, Langton-Hewer R, Wood VA, Skilbeck CE, Ismail HM (1983) The hemiplegic arm after stroke: measurement and recovery. J Neurol, Neurosurg Psychiatry 46(6):521–524
70. Kwakkel G, Kollen B (2007) Predicting improvement in the upper paretic limb after stroke: a longitudinal prospective study. Restor Neurol Neurosci 25(5):453–460
71. Faria-Fortini I, Michaelsen SM, Cassiano JG, Teixeira-Salmela LF (2011) Upper extremity function in stroke subjects: relationships between the international classification of functioning, disability, and health domains. J Hand Ther 24(3):257–265
72. Lo AC, Guarino PD, Richards LG, Haselkorn JK, Wittenberg GF, Federman DG, Ringer RJ, Wagner TH, Krebs HI, Volpe BT et al (2010) Robot-assisted therapy for long-term upper-limb impairment after stroke. N Engl J Med 362(19):1772–1783
73. Kwakkel G, Kollen BJ, Krebs HI (2007) Effects of robot-assisted therapy on upper limb recovery after stroke: a systematic review. Neurorehabilitation Neural Repair
74. Volpe BT, Krebs HI, Hogan N (2001) Is robot-aided sensorimotor training in stroke rehabilitation a realistic option? Curr Opin Neurol 14(6):745–752
75. Masiero S, Celia A, Rosati G, Armani M (2007) Robotic-assisted rehabilitation of the upper limb after acute stroke. Arch Phys Med Rehabil 88(2):142–149
76. Chiri A, Vitiello N, Giovacchini F, Roccella S, Vecchi F, Carrozza MC (2012) Mechatronic design and characterization of the index finger module of a hand exoskeleton for post-stroke rehabilitation. IEEE/ASME Trans Mechatron 17(5):884–894
77. Heo P, Gu GM, Lee S-J, Rhee K, Kim J (2012) Current hand exoskeleton technologies for rehabilitation and assistive engineering. Int J Precis Eng Manuf 13(5):807–824
78. Levin MF, Kleim JA, Wolf SL (2008) What do motor "recovery" and "compensation" mean in patients following stroke? Neurorehabilitation Neural Repair
79. Cirstea MC, Levin MF (2000) Compensatory strategies for reaching in stroke. Brain 123(5):940–953
80. Mji G (2001) International classification of functioning, disability and health

81. Heller A, Wade DT, Wood VA, Sunderland A, Hewer RL, Ward E (1987) Arm function after stroke: measurement and recovery over the first three months. J Neurol, Neurosurg Psychiatry 50(6):714–719
82. ©Arduino. Arduino uno, an open-source electronics prototyping platform. http://arduino.cc/
83. Siciliano B, Sciavicco L, Villani L, Oriolo G (2009) Robotics: modelling, planning and control. Springer, Berlin
84. Brott T, Adams HP, Olinger CP, Marler JR, Barsan WG, Biller J, Spilker J, Holleran R, Eberle R, Hertzberg V (1989) Measurements of acute cerebral infarction: a clinical examination scale. Stroke 20(7):864–870
85. Pan P, Lynch KM, Peshkin MA, Colgate JE (2005) Human interaction with passive assistive robots. In: 9th International conference on rehabilitation robotics, 2005. ICORR 2005, pp 264–268
86. Taub E, Miller NE, Novack TA, Cook EW, Fleming WC, Nepomuceno CS, Connell JS, Crago JE (1993) Technique to improve chronic motor deficit after stroke. Arch Phys Med Rehabil 74(4):347–354
87. Roby-Brami A, Jacobs S, Bennis N, Levin MF (2003) Hand orientation for grasping and arm joint rotation patterns in healthy subjects and hemiparetic stroke patients. Brain Res 969(1):217–229
88. Meng Q, Lee MH (2006) Design issues for assistive robotics for the elderly. Adv Eng Inform 20(2):171–186
89. Stanger CA, Anglin C, Harwin WS, Romilly DP (1994) Devices for assisting manipulation: a summary of user task priorities. IEEE Trans Rehabil Eng 2(4):256–265
90. Miguelez JM, Miguelez MD, Alley RD (2004) Amputations about the shoulder: prosthetic management. Altas of amputations and limb deficiencies-surgical, prosthetic, and rehabilitation principles. American Academy or Orthopaedic Surgeons, Rosemont, pp 263–273
91. Pons JL (2010) Rehabilitation exoskeletal robotics. IEEE Eng Med Biol Mag 29(3):57–63
92. Hussain I, Spagnoletti G, Salvietti G, Prattichizzo D (2017) Toward wearable supernumerary robotic fingers to compensate missing grasping abilities in hemiparetic upper limb. Int J Robot Res 36(13–14):1414–1436
93. Casson AJ, Logesparan L, Rodriguez-Villegas E (2010) An introduction to future truly wearable medical devices-from application to asic. In: 2010 annual international conference of the IEEE engineering in medicine and biology society (EMBC). IEEE, pp 3430–3431
94. Webb J, Xiao ZG, Aschenbrenner KP, Herrnstadt G, Menon C (2012) Towards a portable assistive arm exoskeleton for stroke patient rehabilitation controlled through a brain computer interface. In: 2012 4th IEEE RAS & EMBS international conference on biomedical robotics and biomechatronics (BioRob). IEEE, pp 1299–1304
95. Bicchi A, Bavaro M, Boccadamo G, De Carli D, Filippini R, Grioli G, Piccigallo M, Rosi A, Schiavi R, Sen S, et al (2008) Physical human-robot interaction: dependability, safety, and performance. In: 10th IEEE international workshop on advanced motion control, 2008. AMC'08. IEEE, pp 9–14
96. Konrad P (2005) The abc of emg. A practical introduction to kinesiological electromyography, 1
97. Williams TW III, Altobelli DE et al (2011) Prosthetic sockets stabilized by alternating areas of tissue compression and release. J Rehabil Res Dev 48(6):679
98. Dollar AM, Howe RD (2010) The highly adaptive sdm hand: design and performance evaluation. Int J Robot Res 29(5):585–597
99. Graf C (2008) The lawton instrumental activities of daily living scale. AJN Am J Nurs 108(4):52–62
100. Vanderborght B, Albu-Schäffer A, Bicchi A, Burdet E, Caldwell DG, Carloni R, Catalano M, Eiberger O, Friedl W, Ganesh G et al (2013) Variable impedance actuators: a review. Robot Auton Syst 61(12):1601–1614
101. Manti M, Hassan T, Passetti G, d'Elia N, Cianchetti M, Laschi C (2015) An under-actuated and adaptable soft robotic gripper. In: Biomimetic and biohybrid systems: 4th international conference, living machines 2015, Barcelona, Spain, 28–31 July 2015, Proceedings. Springer International Publishing, Cham, pp 64–74

102. Cecilia L, Matteo C (2014) Soft robotics: new perspectives for robot bodyware and control. Front Bioeng Biotechnol 2(3)
103. Hussain I, Salvietti G, Malvezzi M, Prattichizzo D (2017) On the role of stiffness design for fingertip trajectories of underactuated modular soft hands. In: 2017 IEEE international conference on robotics and automation (ICRA). IEEE, pp 3096–3101
104. Hussain I, Albalasie A, Awad MI, Seneviratne L, Gan D (2018) Modeling, control, and numerical simulations of a novel binary-controlled variable stiffness actuator (bcvsa). Front Robot AI 5:68
105. Awad MI, Hussain I, Gan D, Az-zu'bi A, Stefanini C, Khalaf K, Zweiri Y, Taha T, Dias J, Seneviratne L (2019) Passive discrete variable stiffness joint (pdvsj-ii): modeling, design, characterization, and testing toward passive haptic interface. J Mech Robot 11(1):
106. Birglen L, Lalibertè T, Gosselin C (2008) Underactuated robotic hands. Springer tracts in advanced robotics, vol 40. Springer, Berlin
107. Anwar M, Al Khawli T, Hussain I, Gan D, Renda F (2019) Modeling and prototyping of a soft closed-chain modular gripper. Int J Robot Res Appl, Ind Robot
108. Salvietti G, Iqbal Z, Hussain I, Prattichizzo D, Malvezzi M (2018) The co-gripper: a wireless cooperative gripper for safe human robot interaction. In: 2018 IEEE/RSJ international conference on intelligent robots and systems (IROS). IEEE, pp 4576–4581
109. Hussain I, Renda F, Iqbal Z, Malvezzi M, Salvietti G, Seneviratne L, Gan D, Prattichizzo D (2018) Modeling and prototyping of an underactuated gripper exploiting joint compliance and modularity. IEEE Robot Autom Lett 3(4):2854–2861
110. Hussain I, Al-Ketan O, Renda F, Malvezzi M, Prattichizzo D, Seneviratne L, Abu Al-Rub RK, Gan D (2020) Design and prototyping soft–rigid tendon-driven modular grippers using interpenetrating phase composites materials. Int J Robot Res, pp 0278364920907697
111. Hussain I, Anwar M, Iqbal Z, Muthusamy R, Malvezzi M, Seneviratne L, Gan D, Renda F, Prattichizzo D (2019) Design and prototype of supernumerary robotic finger (srf) inspired by fin ray® effect for patients suffering from sensorimotor hand impairment. In: 2019 2nd IEEE international conference on soft robotics (RoboSoft). IEEE, pp 398–403
112. Eppner C, Brock O (2013) Grasping unknown objects by exploiting shape adaptability and environmental constraints. In: 2013 IEEE/RSJ international conference on intelligent robots and systems (IROS), pp 4000–4006
113. Catalano MG, Grioli G, Farnioli E, Serio A, Piazza C, Bicchi A (2014) Adaptive synergies for the design and control of the pisa/iit softhand. Int J Robot Res 33(5):768–782
114. ©Robotis. Dynamixel mx-28t robot actuator (2012). http://www.trossenrobotics.com/dynamixel-mx-28-robot-actuator.aspx
115. ©ArbotiX. Arbotix-m robocontroller, open source (2012). http://www.trossenrobotics.com/p/arbotix-robot-controller.aspx
116. Gafford J, Ding Y, Harris A, McKenna T, Polygerinos P, Holland D, Moser A, Walsh C (2014) Shape deposition manufacturing of a soft, atraumatic, deployable surgical grasper. J Med Devices 8(3):030927
117. Holland DP, Park EJ, Polygerinos P, Bennett GJ, Walsh CJ (2014) The soft robotics toolkit: shared resources for research and design. Soft Robot 1(3):224–230
118. Odhner LU, Dollar AM (2012) The smooth curvature model: an efficient representation of euler-bernoulli flexures as robot joints. IEEE Trans Robot 28(4):761–772
119. Gopi JA, Nando GB (2014) Modeling of young's modulus of thermoplastic polyurethane and polydimethylsiloxane rubber blends based on phase morphology. Int J, Adv Polym Sci Technol, pp 43–51
120. Malvezzi M, Gioioso G, Salvietti G, Prattichizzo D (2015) Syngrasp: a matlab toolbox for underactuated and compliant hands. IEEE Robot Autom Mag 22(4):52–68
121. Santello M, Flanders M, Soechting JF (1998) Postural hand synergies for tool use. J Neurosci 18(23):10105–10115
122. Çalli B, Walsman A, Singh A, Srinivasa S, Abbeel P, Dollar AM (2015) Benchmarking in manipulation research: the YCB object and model set and benchmarking protocols. ArXiv:abs/1502.03143

123. Falco J, Van Wyk K, Liu S, Carpin S (2015) Grasping the performance: facilitating replicable performance measures via benchmarking and standardized methodologies. IEEE Robot Autom Mag 22(4):125–136
124. Lund AM (2001) Measuring usability with the use questionnaire. Usability Interface 8(2):3–6
125. Demers L, Weiss-Lambrou R, Ska B (2002) The quebec user evaluation of satisfaction with assistive technology (quest 2.0): an overview and recent progress. Technol Disabil 14(3):101–105
126. Pacchierotti C, Salvietti G, Hussain I, Meli L, Prattichizzo D (2016) The hring: a wearable haptic device to avoid occlusions in hand tracking. In: 2016 IEEE haptics symposium (HAPTICS). IEEE, pp 134–139
127. Hussain I, Spagnoletti G, Pacchierotti C, Prattichizzo D (2016) A wearable haptic ring for the control of extra robotic fingers. In: International AsiaHaptics conference. Springer, pp 323–325
128. Massimino MJ, Sheridan TB (1994) Teleoperator performance with varying force and visual feedback. Hum Factors: J Hum Factors Ergon Soc 36(1):145–157
129. Moody L, Baber C, Arvanitis TN (2002) Objective surgical performance evaluation based on haptic feedback. Stud Health Technol Inform, pp 304–310
130. Meli L, Hussain I, Aurilio M, Malvezzi M, O'Malley MK, Prattichizzo D (2018) The hbracelet: a wearable haptic device for the distributed mechanotactile stimulation of the upper limb. IEEE Robot Autom Lett 3(3):2198–2205
131. Prattichizzo D, Chinello F, Pacchierotti C, Malvezzi M (2013) Towards wearability in fingertip haptics: a 3-dof wearable device for cutaneous force feedback. IEEE Trans Haptics 6(4):506–516
132. Meli L, Pacchierotti C, Prattichizzo D (2014) Sensory subtraction in robot-assisted surgery: fingertip skin deformation feedback to ensure safety and improve transparency in bimanual haptic interaction. IEEE Trans Biomed Eng 61(4):1318–1327
133. Schoonmaker RE, Cao CGL (2006) Vibrotactile force feedback system for minimally invasive surgical procedures. Proceedings of IEEE international conference on systems, man, and cybernetics 3:2464–2469
134. Pacchierotti C, Abayazid M, Misra S, Prattichizzo D (2014) Teleoperation of steerable flexible needles by combining kinesthetic and vibratory feedback. IEEE Trans Haptics 7(4):551–556
135. Gescheider GA (2013) Psychophysics: the fundamentals. Psychology Press, London
136. Levitt HCCH (1971) Transformed up-down methods in psychoacoustics. J Acoust Soc Am 49(2B):467–477
137. Arabzadeh E, Clifford CWG, Harris JA (2008) Vision merges with touch in a purely tactile discrimination. Psychol Sci 19(7):635–641
138. Minamizawa K, Fukamachi S, Kajimoto H, Kawakami N, Tachi S (2007) Gravity grabber: wearable haptic display to present virtual mass sensation. In: Proceedings of ACM special interest group on computer graphics and interactive techniques conference, pp 8–es
139. Minamizawa K, Kajimoto H, Kawakami N, Tachi S (2007) A wearable haptic display to present the gravity sensation-preliminary observations and device design. In: Proceedings of world haptics, pp 133–138
140. Schorr SB, Quek ZF, Romano RY, Nisky I, Provancher WR, Okamura AM (2013) Sensory substitution via cutaneous skin stretch feedback. In: Proceedings of IEEE international conference on robotics and automation, pp 2341–2346
141. Guinan AL, Montandon MN, Doxon AJ, Provancher WR (2014) Discrimination thresholds for communicating rotational inertia and torque using differential skin stretch feedback in virtual environments. In: Proceedings of IEEE haptics symposium, pp 277–282
142. Hussain I, Spagnoletti G, Salvietti G, Prattichizzo D (2016) An emg interface for the control of motion and compliance of a supernumerary robotic finger. Front Neurorobotics 10:18
143. Saponas TS, Tan DS, Morris D, Balakrishnan R (2008) Demonstrating the feasibility of using forearm electromyography for muscle-computer interfaces. In: Proceedings of the SIGCHI conference on human factors in computing systems. ACM, pp 515–524

144. Zecca M, Micera S, Carrozza MC, Dario P (2002) Control of multifunctional prosthetic hands by processing the electromyographic signal. Crit Rev ™ Biomed Eng 30(4–6)
145. Merletti R, Botter A, Troiano A, Merlo E, Minetto MA (2009) Technology and instrumentation for detection and conditioning of the surface electromyographic signal: state of the art. Clin Biomech 24(2):122–134
146. Andrea M, Isabella C (2010) Technical aspects of surface electromyography for clinicians. Open Rehabil J 3(1):
147. Farina D, Merletti R (2000) Comparison of algorithms for estimation of emg variables during voluntary isometric contractions. J Electromyogr Kinesiol 10(5):337–349
148. Mohammadreza Asghari Oskoei and Huosheng Hu (2007) Myoelectric control systems - a survey. Biomed Signal Process Control 2(4):275–294
149. Felzer T, Freisleben B (2002) Hawcos: the hands-free wheelchair control system. In: Proceedings of the 5th international ACM conference on assistive technologies. ACM, pp 127–134
150. Platz T, Pinkowski C, van Wijck F, Kim I-H, Di Bella P, Johnson G (2005) Reliability and validity of arm function assessment with standardized guidelines for the fugl-meyer test, action research arm test and box and block test: a multicentre study. Clin Rehabil 19(4):404–411
151. Wolf SL, Lecraw DE, Barton LA, Jann BB (1989) Forced use of hemiplegic upper extremities to reverse the effect of learned nonuse among chronic stroke and head-injured patients. Exp Neurol 104(2):125–132
152. Jakob N (1994) Usability engineering. Elsevier
153. Taylor CL, Schwarz RJ (1955) The anatomy and mechanics of the human hand. Artif Limbs 2(2):22–35
154. Ajoudani A, Tsagarakis NG, Bicchi A (2012) Tele-impedance: teleoperation with impedance regulation using a body-machine interface. Int J Robot Res, pp 0278364912464668
155. Kwakkel G, Kollen BJ, van der Grond J, Prevo AJH (2003) Probability of regaining dexterity in the flaccid upper limb impact of severity of paresis and time since onset in acute stroke. Stroke 34(9):2181–2186
156. Hussain I, Salvietti G, Spagnoletti G, Cioncoloni D, Rossi S, Prattichizzo D (2017) A soft robotic extra-finger and arm support to recover grasp capabilities in chronic stroke patients. In: Wearable robotics: challenges and trends. Springer, pp 57–61
157. Prange GB, Jannink MJA, Groothuis-Oudshoorn CGM, Hermens HJ, IJzerman MJ (2006) Systematic review of the effect of robot-aided therapy on recovery of the hemiparetic arm after stroke. J Rehabil Res Dev 43(2):171
158. Brewer BR, McDowell SK, Worthen-Chaudhari LC (2014) Poststroke upper extremity rehabilitation: a review of robotic systems and clinical results. Topics in stroke rehabilitation
159. Mehrholz J, Hädrich A, Platz T, Kugler J, Pohl M (2012) Electromechanical and robot-assisted arm training for improving generic activities of daily living, arm function, and arm muscle strength after stroke. The Cochrane Library
160. Kwakkel G, Meskers CGM (2014) Effects of robotic therapy of the arm after stroke. Lancet Neurol 13(2):132–133
161. Krabben T, Prange GB, Molier BI, Stienen AHA, Jannink MJA, Buurke JH, Rietman JS (2012) Influence of gravity compensation training on synergistic movement patterns of the upper extremity after stroke, a pilot study. J Neuroeng Rehabil 9(1):1
162. Ates S, Mora-Moreno I, Wessels M, Stienen AHA (2015) Combined active wrist and hand orthosis for home use: lessons learned. In: 2015 IEEE international conference on rehabilitation robotics (ICORR). IEEE, pp 398–403
163. Farrell JF, Hoffman HB, Snyder JL, Giuliani CA, Bohannon RW (2007) Orthotic aided training of the paretic upper limb in chronic stroke: results of a phase 1 trial. NeuroRehabilitation 22(2):99–103
164. Hussain I, Salvietti G, Spagnoletti G, Malvezzi M, Cioncoloni D, Rossi S, Prattichizzo D (2017) A soft supernumerary robotic finger and mobile arm support for grasping compensation and hemiparetic upper limb rehabilitation. Robot Auton Syst 93:1–12
165. Mathiowetz V, Volland G, Kashman N, Weber K (1985) Adult norms for the box and block test of manual dexterity. Am J Occup Ther 39(6):386–391

166. Taub E, Uswatte G, Pidikiti R (1999) Constraint-induced movement therapy: a new family of techniques with broad application to physical rehabilitation-a clinical review. J Rehabil Res Dev 36(3):237
167. Edwards DF, Hahn M, Baum C, Dromerick AW (2006) The impact of mild stroke on meaningful activity and life satisfaction. J Stroke Cerebvasculaar Dis 15(4):151–157
168. Hebb DO (2005) The organization of behavior: a neuropsychological theory. Psychology Press, London
169. French B, Thomas LH, Leathley MJ, Sutton CJ, McAdam J , Forster A, Langhorne P, Price CIM, Walker A, Watkins CL, et al (2007) Repetitive task training for improving functional ability after stroke. The Cochrane Library
170. Kwakkel G (2006) Impact of intensity of practice after stroke: issues for consideration. Disabil Rehabil 28(13–14):823–830
171. Nijland RHM, van Wegen EEH, Harmeling-van der Wel BC, Kwakkel G, Investigators EPOS et al (2010) Presence of finger extension and shoulder abduction within 72 hours after stroke predicts functional recovery early prediction of functional outcome after stroke: the epos cohort study. Stroke 41(4):745–750
172. Veerbeek JM, van Wegen E, van Peppen R, van der Wees PJ, Hendriks E, Rietberg M, Kwakkel G (2014) What is the evidence for physical therapy poststroke? a systematic review and meta-analysis. PloS One 9(2):e87987
173. Van der Lee JH, Wagenaar RC, Lankhorst GJ, Vogelaar TW, Devillé WL, Bouter LM (1999) Forced use of the upper extremity in chronic stroke patients results from a single-blind randomized clinical trial. Stroke 30(11):2369–2375
174. Hussain I, Santarnecchi E, Leo A, Ricciardi E, Rossi S, Prattichizzo D (2017) A magnetic compatible supernumerary robotic finger for functional magnetic resonance imaging (fmri) acquisitions: Device description and preliminary results. In: 2017 international conference on rehabilitation robotics (ICORR). IEEE, pp 1177–1182
175. Vigaru B, Sulzer J, Gassert R (2016) Design and evaluation of a cable-driven fmri-compatible haptic interface to investigate precision grip control. IEEE Trans Haptics 9(1):20–32 Jan

Printed in the United States
by Baker & Taylor Publisher Services